The Story of Our Amazing Universe

Athar Shareef

BSc (Hons) Electrical and Electronics Engineering, University of Manchester

BSc (Hons) Physical Science, Open University

CEng, Fellow IET

Grosvenor House
Publishing Limited

This book is published by
Grosvenor House Publishing Ltd
Link House
140 The Broadway, Tolworth, Surrey, KT6 7HT.
www.grosvenorhousepublishing.co.uk

A CIP record for this book
is available from the British Library

ISBN 978-1-83975-529-3 (Pbk.)
ISBN 978-1-83975-530-9 (Hbk.)

For Aleph and Zhaleh

With the hope that they become as fascinated by
science as their grandfather.

Contents

Acknowledgements

I owe a debt of gratitude to many in bringing this project to a successful conclusion. Firstly, I owe it to my wife, Zuju, for graciously putting up with my endless rhetorical questions and the seemingly never-ending task of putting the book together. I am grateful to her for being at the sharp end of my excited explanations, as I tried to describe a concept, perhaps of quantum mechanics, or explained how we are made of star dust, and for getting excited herself.

I am obliged to members of my globally dispersed family and to my friends who read what I had written, offered suggestions, helped in other ways, or importantly, just encouraged me to continue.

I am particularly grateful to Juliet Amer who so professionally and diligently proof-read the manuscript. My thanks go to Bryn Reade, a photojournalist, who created artwork for use as illustrations in the book. Bryn is also my son-in-law. He lives in Auckland, New Zealand, with my daughter Juhi, a sustainability executive, and their two children, Aleph and Zhaleh.

I am thankful to the reviewers of my early draft: my Indian brothers-in-law, fellow scientist, Sanjar Ali Khan and science enthusiast, Narendra Reddy, and my astronomy-keen Pakistani great-niece, Kulsum Hassan who lives in Melbourne, Australia. They were amongst the first to urge me to publish.

Finally, I must mention two of my good friends in the UK: Robert Pick, whose family has been close to mine for a long time, and Patrick Mannix, in whose company I passed many an enjoyable hour putting the world to right, over pub lunches and walks in the woods. They made me want to write a book they will choose to read end-to-end. I hope I have succeeded!

I am deeply grateful to my alma maters: The University of Manchester, for opening the door to my career, and, as far as this book is concerned, to the Open University.

This book has been achieved because of the OU. They enabled me to study the subjects of my choice - astrophysics, the science of what makes the Universe tick, and cosmology, the science of the formation, evolution, and the likely future of the Universe - at my own pace after I retired. The freedom to choose topics of personal interest on the course was invaluable. Another one of my objectives was to keep my grey cells active after retirement; the OU enabled me to do so in spades. Thanks to the OU I was able to stretch my study over almost 10 years. The flexibility of the OU process, together with the many interesting tutors and students I met on the course, and the quality of the study material and course work provided, produced an environment that we simply must ensure is preserved for the future.

Preface

The power of questions

To question is good. That is how we learn. We question, we search, we find. Never be afraid to ask. Find out why things are the way they are. That is how we progress. I am going to tell you a story that shows the power of questions. It is an amazing story, a story about everything. And the best thing is that it is not make-believe.

Our story came about because lots of people asked questions. They were not satisfied simply being told, "It is so because it is so". They wanted to know why. They thought about it and came up with a likely explanation, called a hypothesis. But they did not say the hypothesis was true. They said it *may* be true.

They then did experiments to test their hypothesis, to see if the explanation always worked. Sometimes it did, sometimes it did not, and sometimes it worked only partly. Where it did not work, they thought some more and tweaked it or changed it for a better explanation. They told others about it. Many of these other people also thought about it and tested it. Some also suggested improvements. Slowly the story came together and was now called a theory.

But the story is never complete. The more we learn, the more we need to find out. The theory continues to be tested. We are always on the look-out for improvements that can be made to it. This is called the scientific method. This is how science works. The people who search for the answers are called scientists.

The study of science teaches many things about life. We learn to be fair, impartial, and truthful. Science tells us not to accept things for granted. Not to believe something merely because someone says it is true, or that 'it has always been so'. That alone does not make it right. Science teaches us always to probe, always to question. You must be prepared to have your own conclusions questioned in turn; to be willing to listen to a counter argument. Perhaps there is new knowledge and evidence that has been discovered, which may invalidate what was previously accepted. Science tells us to be flexible but firm, not to just blow with the wind. Do all that and you will be on the side of the truth.

There are many disciplines in science. Scientists take their names from the discipline they specialise in: for example, geologists focus on geology, and study the structures of the Earth; archaeologists train in archaeology, the study of humankind through the ages based on the evidence they left behind; chemists study chemistry, the properties of chemicals, their composition and how they interact with other chemicals; physicists study the physical nature of matter and energy, the fundamental particles and their properties; astrophysicists are physicists who study the physics of the Universe as a whole and in all its parts, while cosmologists are concerned with the science of how the Universe formed, its evolution and its likely future. The one thing these disciplines have in common is that they all follow the scientific method.

The story I am going to tell you lies primarily in the domains of the astrophysicist and cosmologist. It is about the incredible things that we have discovered about our Universe, and how we achieved this understanding. The Universe is everything that exists: all life, the Earth, the Moon, the planets, the Sun, all the stars we can see, all the others that we cannot, the space between them, all the light and all the energy that fills it.

Our Universe is amazing. It includes the biggest and strangest objects, such as black holes which gobble up everything that comes near them, including other stars and even light itself. But it also includes the tiniest objects, particles such as atoms, electrons and quarks which combine to make every other particle. The world of these tiny particles is a weird and wonderful place, the quantum world, where things can exist in more than one place at the same time, where particles can pass through barriers, and other peculiar things can happen.

Our Universe is big, much bigger than anything you can imagine, much bigger than anything you could possibly see even with the biggest telescope you can build. It is a place in which nothing can ever travel faster than light, and yet one in which it is possible for one twin to grow older faster than the other.

As you go through the book, keep an eye out for the names of the scientists whose work resulted in the discoveries that today allow the story to be told. Read further about them and their achievements. There is a lot you can learn from them. They are the ones who asked the questions, developed the concepts, and did the experiments which today enables us to sit back, read and marvel.

There is much to tell. Come with me and learn about these and other incredible things about our Universe. Everything we know is what we have found through being curious, asking questions, looking for answers, thinking, testing, improving.

Introduction

My qualifications and efforts that paid the bills were in the field of electrical and electronics engineering, which I studied at the University of Manchester, then in IT, and finally in information system consultancy and business. My dreams though went far, far beyond these everyday concerns, into the realms of the wonders of the Universe. I eventually got my opportunity after retirement to study this subject properly at the Open University and grabbed it.

I started thinking about writing a book as a way of recording what I had learnt at the OU, and the pleasure that the process of learning had given me. The book is a way of repaying a debt I suppose. I was well beyond retirement when I completed the degree, did not intend to return to employment, yet did not want all that effort to have benefitted just myself.

Then when my daughter, Juhi, got married in 2011 and presented us with our grandson, Aleph, the next year, I thought such a book would be a neat legacy to leave for him and any other of his siblings that come along. I duly started writing it, coincidentally, on his birthday, September 5, 2012. So, Aleph and Zhaleh (who joined the family in 2015), this book is dedicated to the two of you. But I have arrogance enough to hope that the audience, who will enjoy what I have written, will be wider than just the two of them.

I initially toyed with the idea of a children's book but decided that that would not do justice to the subject. Instead, I set myself a challenge; to write a book that would be informative and educational, relevant for curious young people who have gained some knowledge of science and developed a thirst for learning more about space, as well as for 'grown-ups' who want a better understanding of this fascinating subject. It was important for me that the book should not dumb down the astonishing concepts we have in science but try and explain these in simple terms. I hope I succeed and keep the interest of all the ages. The icing on my cake will be if some of the younger readers are enthused enough to pursue a career in one of the scientific disciplines.

How to get the most out of the book

I have laid out the topics in the book as generally progressing sequentially in time, from the birth of the Universe, through its evolution and maturity, to where it is today. I also peer into the future and consider its old age and death, or whatever else may be its end, and along the way review the theories we have developed and try to reason why.

During this journey, I discuss several specific topics: the stars, what they are, how they are born, live their lives and die; the extreme objects in space, the ones I have called 'the beasts', the black holes, neutron stars and their like; our solar system, its birth and expected end; the possibility of extraterrestrial life; and the future of the Universe. Each topic has a Chapter devoted to it with an introductory section preceding the main text. Finally, there are a few necessarily complex discussions which I have highlighted, which can be skipped without loss of the thread of the narrative and saved perhaps for a later visit.

My aim, to try and explain each concept in simple terms, was quite a goal to take on. Only you can judge how well, if at all, I have achieved it. There is virtually no mathematics in the book. I have explained notations such as those needed to deal with extreme numbers. Mathematicians developed these to make their own lives easier, so hopefully, they will be helpful for you too.

I have used capital letters for words that relate to specific objects rather than to the general. Hence, Earth instead of earth and Sun instead of sun, refer to our home planet and our own star, and not to the multitude of other planets and stars in the universe.

NASA has an extensive library of space images and articles, including for young children, which is well worth browsing for itself. The internet is a mine of information. You should use it not only for the images, but also to keep up with the latest developments in this fast-moving field. But you also must be careful of the internet. People sometimes post incorrect, or malicious information, and sometimes the information can be transient. If you keep to the reputable sites though, you should not have a problem.

Copyrights

There are numerous illustrations and images in the book. The illustrations are my work. A few images have been provided freely to me. The many digital images are courtesy of external organisations, particularly NASA and its associates. NASA directed by USA copyright law, and some other organisations such as ESA, the European Space Agency, follow an enlightened policy under which some of the images they own are made freely available, provided they are appropriately credited and used. Other images have been obtained under "Creative Commons Licences" that, with suitable attribution, permit the free use of otherwise copyrighted work. A few have been acquired from online image providers.

The credits for all images are given in a comprehensive Image Credits list at the end of the book, rather than with the captions to the images themselves. Each entry is cross-referenced to the image it refers to by the Fig. number given. I have endeavoured to acknowledge and credit each source as requested. Where appropriate, I have provided the image website URL. If any errors or shortcomings should be found, they will be unintentional. I will be grateful to be notified so that they may be suitably rectified.

Chapter 1

In the beginning…

the Big Bang, and how we know it happened

In the beginning there was nothing. Not a sausage. Zilch. No stars, no light, nothing. It was not empty space, because there was no such thing as space, it had not yet been created. There was nothing, not even time. Can you imagine nothing? It is not like standing and looking at an empty box. There was no box. There was no you to be looking at it. Strange, isn't it?

Then something happened. An enormous, humongous, ginormous explosion happened. Except it was not an explosion that you could stand by and watch. It was space and time being created. Why? How? What was there before the 'explosion'? We do not yet know. Remember that even the best theories can still have gaps in them. This is one of those gaps, a big one admittedly. But there are ideas around which scientists think may explain things, and we shall cover some of these briefly later in the book. However, these are not yet accepted as part of the theory. It could be that soon one of the ideas gets tested and accepted as the correct explanation. Or it could be that someone (someone like one of my younger readers perhaps), will come up with another explanation in the future. Till then we have philosophy, belief, hypotheses, and ideas.

But we are sure the 'explosion' happened. We have even given it a name. We call it "The Big Bang".

At this point you should be asking: "But, how do we know it happened?" I will tell you. That is what this chapter is all about. Along the way we will talk about many things: about space, matter, light, stars and galaxies, atoms and, yes, quarks; about how we know what the stars are made of and how we can study the past. You will see why we need to know these things to know how the big bang happened. This is the way science progresses. As we learn about things, each new piece of knowledge helps us to answer a little bit more of the puzzle.

Let us start by finding out what is in our Universe.

Space, without which there is nothing

The thing that fills up the Universe is space. You can think of space as being the surface of a balloon that is our Universe. That makes for a useful analogy, as we shall see. Space is altogether a very strange place. When we look out into the night sky, we can see that there are planets and stars which are floating in space. With a powerful telescope we can also see dust and gas clouds and other galaxies. And we can see light. But even in places where we can see nothing and space is a vacuum, it is not empty. It is full of strange things such as *dark matter* and *dark energy* that we know little about. Space is a place where tiny particles suddenly pop into existence for a smidgen of a second and then disappear as quickly in a puff. Space is where everything exists, the stage where all the action happens.

We have always been fascinated by space. We say the rocket went up in space. We call it the final frontier. We talk of spaceships and imagine going to distant galaxies by travelling through it. Without space there would be no story, nowhere for anything to exist, including us. We shall talk more about space later. We shall see that space is not static but growing bigger even as we speak. The energy in it is pushing against gravity and winning; new space is being created and the Universe is getting bigger.

Once upon a time there was no space and nothing existed, until it was all created by the Big Bang. What happened? How did it begin? How has it changed? What will become of it and all the stuff in it? That is what this book is about.

Let us next look at what things exist in space. We will talk about matter, the stuff that things are made of, and we will talk about light. Then we will see how we can pull together all that we learn to answer the question "How do we know the Big Bang happened?"

Matter, the stuff that things are made of

If you look around you will find things that you can see, hold, touch, or feel, things like stones, leaves, air, and water. These things are all made of stuff we call *matter*. Matter is something that occupies space. *Mass* is a measure of how much matter there is in an object. We can say that the Earth has more mass (that is, it is larger being made of more matter and is therefore more *massive*) than the Moon.

Weight is the force with which the mass of one object, pulls on the mass of another. On Earth, your weight is the force with which the Earth pulls on the mass in your body.

You yourself are made of matter and have mass, and so have weight on Earth; you can measure how much on your bathroom scales [Fig. 1.1].

Chapter 1. In the beginning...

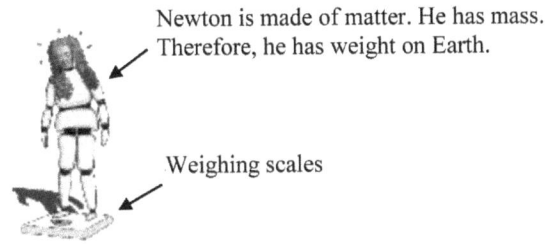

Newton is made of matter. He has mass.
Therefore, he has weight on Earth.

Weighing scales

Fig. 1.1 Mass, Matter and Weight.

Gravity is the force with which mass attracts other mass. Your weight is the force due to gravity with which the mass of the Earth attracts the mass of your body, and equally in turn, your body's mass attracts the Earth's. The more massive the objects, the larger is the force due to gravity (or gravitational force) between them. How much things weigh depends on the value of gravity where they are. For example, you probably know that you will weigh much less on the Moon than here on Earth. As the Moon is much less massive than the Earth, it attracts you with a smaller force than does the Earth. We say that the gravitational force on the Moon is less than that on the Earth. In fact, the gravity on the Moon is one-sixth of that on Earth: you will weigh about six times less on the Moon than you do on Earth.

Gravity is an incredibly important force of nature. Because mass attracts other mass, the bodies with mass, such as the Sun, its planets, the Earth and Moon, and people, are all attracted to each other. This is the reason why the planets go in an orbit around the Sun, why the Moon goes around the Earth, why we stay on the ground, and why all these objects do not fly away from each other. Yet gravity is a very weak force. Look at it this way: an apple that breaks off a branch falls because the whole mass of the Earth is pulling it towards its centre; it stops only when the ground gets in the way.

You can however simply bend down and pick this apple up. The force of gravity due to the whole of the Earth is unable to stop you from doing so. The power in your muscles has easily overcome the force of attraction due to the whole Earth.

Speaking of apples, I am sure you have heard about Sir Isaac Newton (1643-1727), one of the great genius scientists to whom we owe so much. A story is told that Newton was sitting under an apple tree when one of the apples fell (we do not know whether this landed on his head or the ground) [Fig. 1.2].

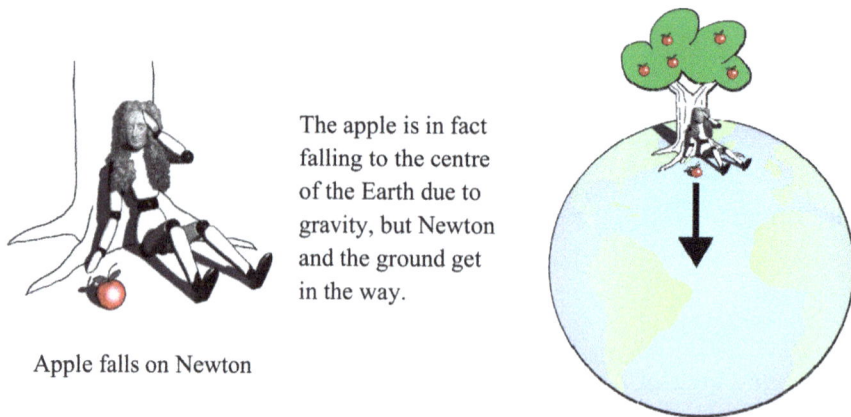

The apple is in fact falling to the centre of the Earth due to gravity, but Newton and the ground get in the way.

Apple falls on Newton

Fig. 1.2 Newton and the apple.

He questioned why this happened, why the apple did not fly off into space instead. He realised that there must be a force attracting the apple to the ground and worked out the law that showed how objects are attracted by gravity. He also produced laws that describe how bodies move when forces act on them. We use Newton's laws of motion to send rockets into space.

Newton realised that the reason why gravity appears so strong is that it increases with the mass of the objects on which it acts. Therefore, a little boy weighs less than a sumo wrestler. And why when gravity acts between very massive objects like the stars and the planets the force can be huge. The force due to gravity also depends on how near the objects are to each other [Fig. 1.3]. The nearer these objects are to each other the stronger is the gravitational force attracting them to each other.

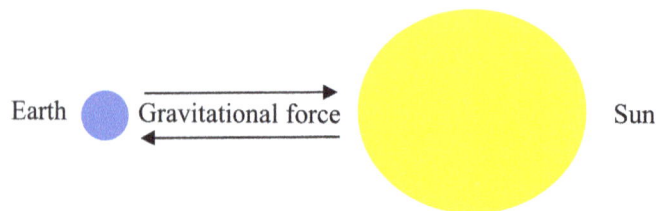

Earth   Gravitational force   Sun

- Gravity attracts each body to the other.
- The nearer the bodies are to each other, the greater is the force between them.
- The more massive the bodies are, the greater is the force due to gravity.

Fig. 1.3 Gravity attracts all things that have mass.

Chapter 1. In the beginning…

Newton worked out that the force due to gravity not only makes apples drop to the ground but also keeps all the planets in the solar system going round the Sun. Later, when we talk about black holes, we will see how big the force due to gravity can really get.

We have seen that one of the things that make up our Universe is matter, which is stuff that has mass. Matter is affected by the gravitational forces due to all the other matter which exists in the Universe.

Some of the matter in the Universe is in the form of stars, some of these we know have planets orbiting around them, each also made of matter. These stars together with their planets are further grouped into huge galaxies, which can have some 200 billion stars (that is two hundred thousand million, a large number which we write as 200,000,000,000). There are estimated to be some 200 billion galaxies in the observable Universe. This is an awful lot of stars, made up of an awful lot of matter. Someone worked out that there are more stars in the Universe than all the grains of sand on all the beaches of the world. But there is even more matter which is not in stars and planets and which exists in the dust and gas floating in space within and between the galaxies.

By now you would have gathered that the Universe is BIG and has a lot of matter in it. All this matter in space is kept in place by the attraction of all the other matter due to gravity.

The scientists however have a problem. When they calculated the total amount of matter that can be seen in the Universe and worked out the total amount of matter that is needed for gravity to keep the stars and galaxies moving around each other without flying off into space, they were shocked to find that the matter that we can see was less than 15% of the matter that is required. In other words, some five or more times of additional matter was needed to stop the galaxies flying apart. Where is this extra matter that we cannot see? What is it?

Well, embarrassingly, we do not know. We have some ideas but extraordinarily little proof about what makes up the missing matter. For this reason, we call it *dark matter*. Many scientists are working to solve the mystery. But for now, let us say that this is another of those gaps in our knowledge that will be up to the scientists of the future to solve. We will return to dark matter later.

Light that we see, and light that we do not see

Something else that we can see when we look around our Universe is *light*. In fact, the reason why we see anything is because of light. Light bounces off objects and gets into your eyes where it is focussed on the retina at the back of your eyeball. The cells in the retina send signals to your brain about the light that has been focussed. Your eyes [Fig. 1.4] enable you to see the object. You should note that the eyes of some other creatures are quite different in structure from ours and see things very differently to us humans.

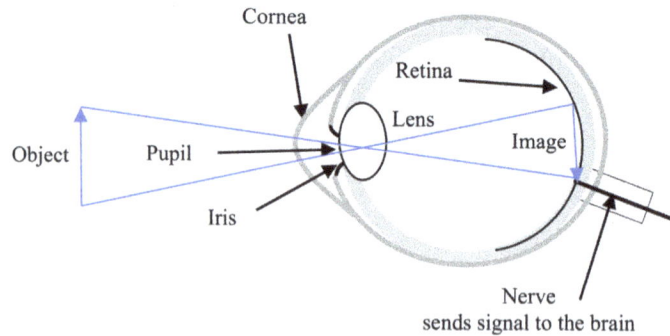

Fig. 1.4 The structure of the human eye (much simplified).

Light is not matter. It has no mass and therefore no weight at all, wherever in the Universe it may be. It has no weight on the Earth or on the Moon, not on any of the stars nor anywhere else in space. Light is in fact energy in the form of *electromagnetic radiation*. I will try and explain.

You should already know about electricity which we use to light up our homes, and heat the water for tea, or, in my case, coffee. You would also know about magnetism; I am sure that you have seen magnets that can attract metal and pick-up paper clips. A very clever Scottish scientist called James Clerk Maxwell (1831-1879) used mathematics to analyse electricity and magnetism. He showed that light, electricity, and magnetism were in fact part of the same thing called *electromagnetism*. He proved that light was made up of electrical and magnetic energy that travelled as *electromagnetic* waves.

Maxwell calculated the speed of light and showed that it moved extremely fast indeed: about 300,000 (three hundred thousand) kilometres/second, or about 186,000 miles/second. This means that in one second light can go a distance that is more than 7 times round the Earth; it can also travel to the Moon which is about 384,400 km or 240,000 miles from us in approximately 1.3 seconds. Moreover, Maxwell showed that light's speed was constant at this value when it was travelling through a vacuum such as space. [Important note: Light's speed depends on the medium through which it is travelling. When light travels through a denser medium such as glass or water, its frequency remains constant, but its wavelengths decrease, and as we shall see later, this means its speed slows down. Light travels the fastest through a vacuum.]

Maxwell's work was particularly important for our understanding of how the Universe works.

The light we can see is only a small part of all the electromagnetic radiation (energy) that is all around us. Many creatures can see parts of the spectrum (the range of light colours) that are invisible to us. For example, some insects can see colours in the flowers that shine in ultra-violet light.

X-rays, which we use to see inside a body to check for broken bones, are also electromagnetic radiation, as are ultra-violet rays that give us sunburn, microwaves we use to heat up our food, radio waves we use to broadcast music and other programmes, and TV signals that bring pictures into our homes. Even the heat that comes to us from the Sun is a form of electromagnetic radiation that we call infra-red radiation. In science, the word 'light' is often used for all types of electromagnetic radiation.

But what is the difference between visible light we can see and these other forms of light that we can sometimes feel, like heat, but cannot see? Let me tell you.

When you have been down to the seaside you would have seen the waves come rolling onto the shore. Sometimes the waves are big, especially if the wind is blowing, while at other times they are gentle ripples. Each complete up-down-up movement of a wave is called its *cycle* (or *period*) and the number of cycles passing by in a second is called the *frequency* of the wave [Fig. 1.5 (a)].

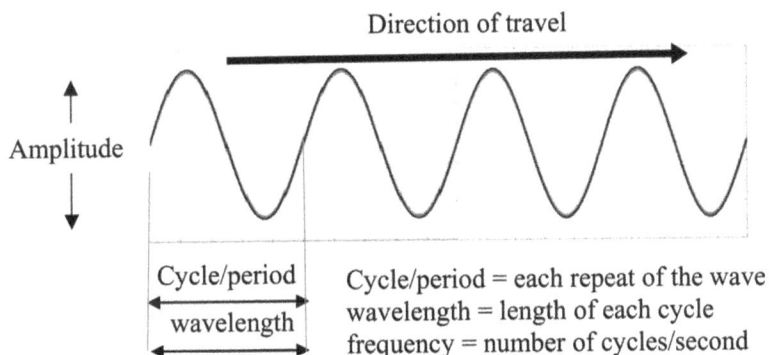

Direction of travel

Amplitude

Cycle/period
wavelength

Cycle/period = each repeat of the wave
wavelength = length of each cycle
frequency = number of cycles/second

Fig. 1.5 (a) Travelling wave.

You will see that if you were to take a wave and stretch it out, it will have a lower frequency, but it will be longer [Fig. 1.5 (b)]. The length of a wave is called its *wavelength*. If you stand in one place and watch the waves, you will see that the longer the wavelength of a wave the fewer waves will go by in any given time and therefore the frequency of the wave will be lower. Similarly, the shorter the wavelength of a wave, the higher will be its frequency. You can check from the calculations given with the diagram below that the speed of a wave is its wavelength multiplied by its frequency. The height of a wave is called its *amplitude*; but this is not of relevance for this discussion.

Distance travelled in 1 second

Wave has a frequency of 2 cycles per second

Wavelength = 10 cm

2 waves go by each second

Each wave is 10 cm long.

Wave travels 10 cm x 2 = 20 cm in 1 second

10 cm

Speed of the wave is 20 cm per second

Wave has a frequency of 4 cycles per second

Wavelength = 5 cm

4 waves go by each second

Each wave is 5 cm long

Wave travels 5 cm x 4 = 20 cm in 1 second

5 cm

Speed of this wave is also 20 cm per second

(NB Physical waves, such as sea waves or sound waves, do not always have the same speed)

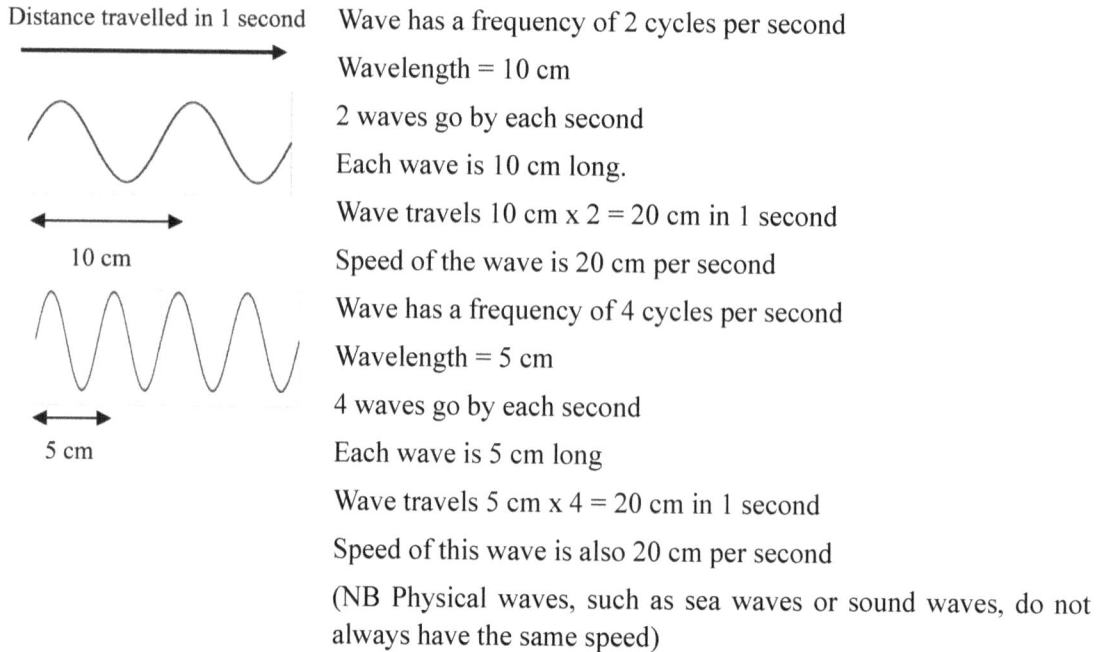

Fig. 1.5 (b) Frequency and wavelength of travelling waves.

Each of the types of electromagnetic radiation I mentioned above - radio waves, microwaves, heat, visible light, x-rays, and so on - are the same thing, except for their frequency and wavelength. Some waves such as x-rays go up and down (*oscillate*) many more times every second, that is they have a higher frequency (and therefore smaller wavelengths) than others such as radio waves. Light with a higher frequency has more energy than light with a lower frequency. Thus, x-rays have more energy than visible light, and so can penetrate deep into our bodies.

Because x-rays have more energy than most other radiation, they can also do more harm by damaging the cells in our bodies. That is why we are careful when we use x-rays, and house them in protected rooms and put a warning light outside.

But whatever their frequency and wavelength, *all electromagnetic waves travel with the same speed in a vacuum*, which is the speed of light. Air is a good approximation of a vacuum for this purpose. Physical waves, such as sound waves or waves in the sea, or pond, do not all have the same speed.

The retina at the back of our eyes is sensitive only to a certain range of light frequencies. In other words, our retina sends signals to our brain only when light of these frequencies strikes it. This is the part of the light that we can see and which we call visible light.

High frequency
High energy
(Violet)

Low frequency
Low energy
(Red)

(a) Visible spectrum of light from the Sun as seen in a rainbow or a prism.

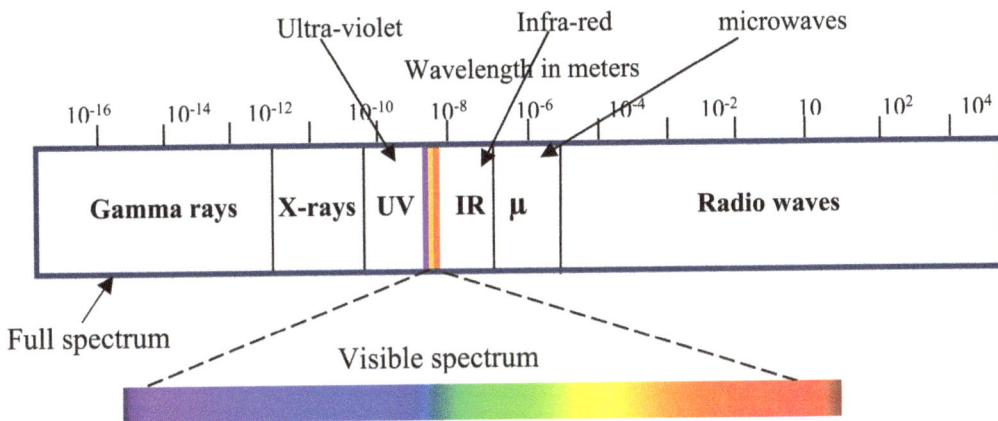

Ultra-violet   Infra-red   microwaves

Wavelength in meters

10^{-16} 10^{-14} 10^{-12} 10^{-10} 10^{-8} 10^{-6} 10^{-4} 10^{-2} 10 10^2 10^4

| Gamma rays | X-rays | UV | IR | μ | Radio waves |

Full spectrum

Visible spectrum

(b) The full electromagnetic spectrum, showing the visible spectrum.

Fig. 1.6 The electromagnetic spectrum.

The remarkable thing about visible light in the world around us is that we can see colours. You will of course have noticed this as you gaze around you. You would also have seen the colours in rainbows after a rain shower, or when sunlight is passed through a prism. Sunlight is made up of all these colours. The spread of colours we see is called a visible spectrum [Fig. 1.6 (a)]. Each colour occurs at its specific frequency. Remember there are other frequencies on either side of the visible light that we cannot see [Fig. 1.6 (b)]. The visible spectrum that we see is only a small part of the full electromagnetic spectrum.

In a spectrum produced by a rainbow or by using a prism, the colours we see range from violet at one end, through indigo, blue, green, yellow, and orange to red at the other end. [Can you remember this sequence? I do it by remembering the made-up word VIBGYOR, where V is for violet, I for indigo and so on. Some remember it by making up a sentence such as "**Richard Of**

York **G**ave **B**attle **I**n **V**ain" which gives us the colours in reverse order to my made-up word. You can try to figure out your own way to remember].

The sequence of the colours is important. We have seen that the only difference between the different parts of the spectrum is the frequency of the light that we can see. Violet has the highest frequency in the spectrum. Red at the other end of the spectrum has the lowest frequency. Because of its higher frequency, you will have realised that violet light has more energy than red.

Violet light is the highest frequency light we can see. Beyond it, at a higher frequency than violet light, is ultra-violet (UV) light. This light gives us sunburn and can damage our skins; we protect ourselves against it by using sun lotions. Red light is the lowest frequency (and longest wavelength) light we can see. Beyond it, at a lower frequency, is infra-red (IR) radiation; this is the frequency at which heat is transported to us from the Sun.

If you pass sunlight through a prism and put a thermometer in the infra-red part (just beyond the red band) of the spectrum that is produced, you will find it reads a higher temperature than if it is in, say, the ultra-violet part (just before the violet band) at the other end of the spectrum.

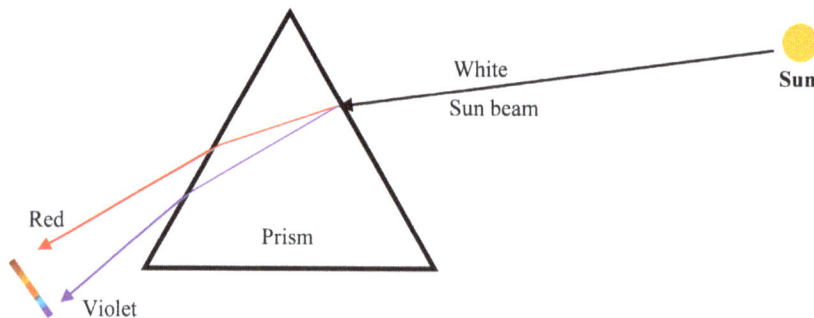

Fig. 1.7 How the spectrum colours are produced by a prism.

A prism separates the colours of the white sunlight because the light of different frequencies bends (*refracts*) through different angles when going across two different media: from a less dense medium (air) to a denser medium (glass), or the other way from glass to air [Fig. 1.7].

We have noted earlier that the speed of light depends on the medium it is travelling through, such as air, glass, or water. Its frequency stays constant in all media, but the wavelengths get shorter in a denser medium. The higher frequency, shorter wavelengths reduce more than those of lower frequency. Going from air to glass, violet light with its high frequency and short wavelength slows down and refracts the most, while red light with its large wavelength refracts the least. Thus, in glass, the lower wavelength violet light is bent more than red light with the other frequencies in between, resulting in a spectrum.

Chapter 1. In the beginning...

Refraction is also how a rainbow is produced, the water raindrops in the atmosphere doing the same job as a prism.

This bending and scattering of light also explain why the sky is blue during the day, and why the sunsets have the reddish orange colours. The sunlight that streams to us from the Sun gets scattered by the gases and dust particles in our atmosphere. The high-frequency blue light is scattered more, and more of it enters our eyes from all parts of the sky than the other colours. That is why the sky looks blue. (In fact, violet is a higher frequency than blue, and is scattered the most. However, the sunlight has a higher density of blue light, and our eyes are more sensitive to blue than to violet. That is why the sky is blue and not violet.) When the Sun is shining directly at us, all colours reach our eyes equally and the light looks white. At sunset, when the Sun is lower in the sky, it is the blue/violet light that is scattered away from our eyes and the other more-red colours get into our eyes from the setting Sun giving us the red and orange sunset colours.

One of the scientists who investigated the spectrum of visible light was Sir Isaac Newton, about whom we spoke earlier. Newton used a prism to split the Sun's light into the rainbow colours. He was the first person to study the spectrum in detail, and to understand it. He also invented the reflecting telescope. Newton really was a genius.

How can light tell us what stars are made of?

Today, we can observe the Universe through powerful telescopes. We can look far into the distance with them. Most of the telescopes are on Earth, many sited on the tops of mountains above the clouds and in deserts with cloudless skies so we can get clearer images. Some telescopes we have put into orbit to get above our shimmering atmosphere altogether. Telescopes are essential tools in helping us solve many of the questions about our Universe. To see the part that telescopes have played in our story, I need to tell you some interesting things about them.

A telescope is made up of lenses and mirrors which focus light from distant objects in such a way that the images are magnified, that is the objects look bigger and nearer. Perhaps you yourself have seen through a telescope, or a pair of binoculars. We will not go into how a telescope works in this book. Here I want to tell you how telescopes have helped us know how the Big Bang happened.

You will see as we go along that we do know a lot about stars. We know these things because we can see the stars through telescopes. But that is all we can do: look at them. We cannot visit them as they are too far away. We cannot send a rocket to collect a bit of the star and bring it back to look at on Earth, because the rocket will take too long and, in any case, it will burn up as it approaches the star. No, the *only* thing we can do to study stars is look at them through a telescope and examine the light that is coming to us from them. I find that interesting. I hope you do too.

Let me give you an example of how we can use light to tell us about the Universe. But first we need to digress briefly to talk about atoms and molecules. Later we will talk a lot more about atoms and see the strange world they occupy.

Earlier we spoke about matter that is everywhere in our Universe in the form of stars, planets, galaxies, dust and gas, seas, trees, you, and me. This matter is made up of extremely tiny things we call atoms. Each atom [Fig. 1.8 (a)] has a nucleus in the centre containing particles called protons and neutrons; even tinier bits of matter called electrons whirl around this nucleus.

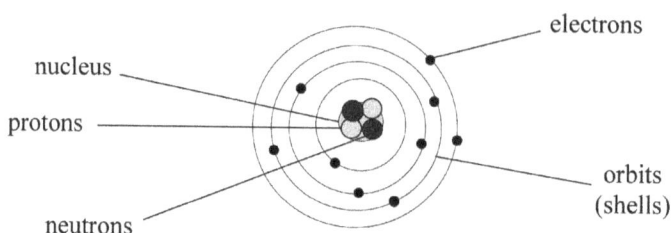

(a) The structure of atoms.

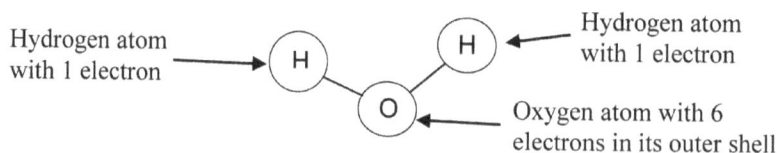

(b) The structure of the water molecule.

Fig. 1.8 The structure of atoms and molecules.

The electrons spin around the nucleus in fixed orbits (also called shells). Protons and neutrons are much bigger than the electrons. In stable atoms the total number of electrons equals the number of protons (NB this is not shown in the diagram for clarity). The numbers of these particles determine the element of the atom. For example, a hydrogen atom has one proton and helium has two protons and two neutrons in their respective nuclei.

Rules of nature dictate the numbers of electrons in each shell (again, this rule is not shown in the diagram). The outermost shell is particularly important. Stable atoms have 8 electrons in this shell. Atoms that do not, look for partner atoms with whom they can share electrons to make up this number. This combination of stable atoms is called a molecule.

Chapter 1. In the beginning…

Thus, an oxygen atom which has 6 electrons in its outermost shell can team up with two hydrogen atoms each of which has one electron. This results in 8 shared electrons in each atom's outer shell, which gives a very stable molecule that we write as H_2O and we call water [Fig. 1.8 (b)].

Now back to our story.

We have seen that apart from matter the other thing that fills the Universe is light. In its journey through space, light comes across matter and sometimes they even collide. What happens then?

When light hits an atom or a molecule, what happens depends on the frequency of the light and the type of atom or molecule it strikes. We have seen that matter is made up of atoms of different substances. There are over a hundred basic substances in nature that cannot be broken down into any other simpler substance; these are called elements. Examples of elements include hydrogen, oxygen, carbon, copper, iron, nickel, gold, etc. Atoms of two or more elements can combine to create molecules of new substances, such as an atom of oxygen which combines with two atoms of hydrogen to produce a molecule of water, as we have seen.

Atoms and molecules will absorb light of frequencies special to them but scatter all other light. So, a hydrogen atom will always absorb light of certain specific frequencies, while another atom, say that of carbon, will absorb light of different frequencies. In its journey to us from a distant star, the light beam from the star will encounter the atoms and molecules of the star itself as well as of other substance it meets along the way. Thus, each time that light collides with atoms and molecules of matter, light of specific frequencies will be removed from the beam depending on the atoms and molecules it met.

Now comes the interesting bit. Sitting here on Earth with just a telescope, we can examine the spectrum of the light by passing it through a prism. [This study is called spectroscopy]. Sometimes, we see black lines in the spectrum [Fig. 1.9].

Fig. 1.9 Spectral lines in the spectrum of light from a star.

Because the lines are black, we know that there is no light at those frequencies in the spectrum. The dark lines show the light that has been 'lost' (or absorbed) along the way. If the black lines occur at the special frequencies which are absorbed by hydrogen atoms, we know that our light beam met hydrogen atoms along its journey. Similarly, black lines which are seen at the carbon frequencies tell us that the beam came across carbon atoms.

Since the light beam is coming from a star, it would have passed through the gases that make up the star; it would also have passed through gas and dust clouds on its journey to Earth. Thus, this spectrum may tell us, for example, that the stellar (belonging to a star) gas atoms that the light met along the way were hydrogen and carbon, and the dust clouds contained silicon and iron among many other elements and compounds. (NB a *compound* is a substance containing the atoms of two or more elements that are *chemically bound together*. A molecule of water is a compound. A *mixture* of salt and sugar molecules is not). So, sitting here on Earth, just by examining the light, we can tell that the star is made up of at least hydrogen and carbon and that the dust and gas clouds contained silicon and iron compounds. Each of the elements has its own light signature made up of the light frequencies that it absorbs. This is how just by studying the spectrum of the light from distant space we can tell the compositions of far distant objects. Imagine that.

<u>Interesting experiment</u>: Pass white (all frequencies) light through a gas and note the black lines in its spectrum. Next, heat the same gas and again note its spectrum without passing any light through it. We will find that the second spectrum (called an emission spectrum) formed by light from the heated gas will have lines *of the appropriate colour* at the exact same frequencies as the black lines in the first spectrum. The heated gas is releasing the light it absorbed!

Let us try and understand light further, starting with its speed.

The speed of light in space is always, always the same

Earlier I mentioned that nothing can travel faster than light in a vacuum. This is a law of nature. That means it is true everywhere. It was proven by the work of a few very clever scientists. One was Maxwell whom we mentioned earlier. Another is one you have almost certainly heard of: Albert Einstein (1879-1955). Here we will look at the role that Einstein played in developing the physics of light. We will be talking much more about him later.

Einstein was a theoretical physicist. This means that he was not the person who would conduct experiments in a lab. He was good at thinking. So, he would think deeply about a problem and come up with theories about the likely solution to the problem. Others then conducted experiments to see whether his theories were right. So far, everything that has been tested about his theories has proven to be correct. Einstein was very bright!!

Chapter 1. In the beginning…

Einstein would do what he called 'thought experiments'. For example, he would imagine what would happen if he could ride on a beam of light. We will talk more about Einstein's thought experiments later. But here we are interested in the fact that he built on Maxwell's work.

Fig. 1.10 Albert Einstein.

Maxwell had shown that the light was a propagating wave of electrical and magnetic fields which travels at a constant speed. Since it is a wave, light can never stand still. Einstein showed that the speed of light is not just constant in a special place or at a special time, but is *always*, *always* the same (in the same medium, of course). Now I mean something interesting by this.

Imagine you are in a car on a motorway, and that in the next lane there is another car travelling at the same speed and direction as yours. If you look across at the other car (very, very carefully if you are the driver!) it will seem as if it is not moving at all: you can look inside the car; see who is there; if someone is reading a newspaper in the car, you may even be able to read what the headline says. We say that as seen from your car, or *relative* to your car, the other car is not moving. However, if the other car was on the other side of the motorway going in the opposite direction, it would whizz by you as a blur. In this case *relative* to you (that is, as seen by you) the other car is travelling faster than its actual speed; in fact, it will seem to be travelling at its own speed plus your own.

However, if instead of a car, you were in a spaceship travelling at high speed and you measured the speed of a beam of light moving by you, the beam you see *will always be travelling at the speed of light relative to you, whether it was moving towards you or away from you, whatever your own speed may be*.

This strange phenomenon of the speed of light was shown by Einstein, building on Maxwell's work, to be the case in his famous *theory of special relativity*. We will talk more about light and its behaviour in Appendix A when we consider Einstein's theories.

Seeing the past

At its speed, light can get to the Moon in approximately 1.3 (just over one and a quarter) seconds, and to the Sun in about 8 minutes. We say that the Moon is 1.3 light seconds, and the Sun is 8 light minutes from us in distance. What this means is that when an astronaut on the Moon waves his or her hands towards the Earth, and we look at them through a powerful telescope, the light from the astronaut's hand will take 1.3 seconds to reach our eyes. So, we will be seeing the astronaut as they were 1.3 seconds ago. In other words, we will be seeing the astronaut waving 1.3 seconds *in the past.*

Our nearest neighbouring star Proxima Centauri is 4 light years from us. Its light therefore takes 4 years to reach us, so we see this star as it was 4 years ago.

The further away the star is the longer its light takes to reach us and the further into the past we are looking. *A telescope is like a time machine that enables us to probe the past of our Universe*!

But we still have not said how this tells us that there was such a thing as a Big Bang. To do that, we must introduce another of the pioneers of science. This time it is an American astronomer called Edwin Hubble (1889-1953). We are going to use the speed of light and its spectrum to show that the Universe is expanding and then to deduce that there was a Big Bang!

Stay with me.

The Universe is 'like a balloon'

You may have seen some of the amazing photos taken by the Hubble telescope that NASA sent into space in 1990 (if not, go onto the internet and Google *Hubble telescope images* and look at them, or try the NASA site). Edwin Hubble was such an important astronomer that NASA named the telescope after him. He was an observer who spent his time looking at the stars to measure how far they are. There are several ways we can find out how far the stars are, but we will talk about these later. Hubble was able to identify stars and galaxies and to put a distance to them. He measured this distances in light years.

Remember we said that light from the Sun takes 8 minutes to reach us, and so the Sun is 8 light minutes away from us. The objects that Hubble was measuring were so far away that their light took millions of years to reach us. We can therefore say that these objects are millions of light years from us.

A light year is a long, long way. Light travels 300,000 kilometres per second, so in a year it would travel approximately 10^{13} kilometres (that is, 1 with 13 zeros after it), or 10^{16} metres in a year. We use metres and kilometres in science, but in our every-day units in the UK we use miles. You can easily convert kilometres to miles by multiplying by 5 and dividing the result by 8. Thus, one light year is about 6 million, million (6 trillion) miles. You could also write this as 6×10^{12} or 6,000,000,000,000 miles. So, you see a light year is an exceedingly long way indeed. As we mentioned earlier, Proxima Centauri, our nearest star is 4 light years, or 24 trillion miles, from us. The objects that Hubble was looking at, such as the Andromeda [Fig. 1.11] the nearest major galaxy to us, were millions of light years away. He was seeing the Andromeda galaxy, for example, as it was almost two and a half million years in the past. Now that is proper time travel.

Incidentally, an interesting fact about Andromeda is that it is part of 'nearby' group of galaxies called the Local Group to which our galaxy, the Milky Way, also belongs. This group is linked by the gravitational attraction between its member galaxies. Its galaxies move independent of other more distant galaxies. Andromeda happens to be moving towards us. In some 4 billion years it will collide with the Milky Way galaxy. More on this later.

Fig. 1.11 The Andromeda galaxy.

After he measured how far the objects were in his telescope Hubble looked at the spectrum of their light. He saw that the spectrum had many black lines across it showing that the light had met many different elements in its journey over the vast distances to us. But then he noticed that the lines in the spectrum were not in their expected places. No matter in which direction he looked at objects in the sky, the frequencies of the lines were different to what was anticipated [Fig. 1.12 (a)]. It was as if the atoms were absorbing light of wrong frequencies. This was very peculiar indeed, since all hydrogen atoms will always absorb light of the same frequencies, as will other atoms at their own special frequencies. Hubble reasoned that if that fact was correct (and that it was correct had been proven by experiments), then somehow the frequencies were being changed as the light came to us.

Hubble had a flash of inspiration. He saw that if he moved the spectrum from the different galaxies across each other the frequencies could be lined up [Fig. 1.12 (b)].

He also noticed that the further the galaxies were, the further he had to move the spectrum. In other words, Hubble found that the further that the galaxies were the more their frequencies had changed. He also saw that generally the frequencies had become lower (that is the wavelengths had become longer) and the spectrum had shifted towards the red end. It was as if the light waves were being stretched, and the further the objects were, the greater was the stretching. He was helped by the efforts of an American astronomer Henrietta Leavitt (1868-1921) which enabled the distances to the far galaxies to be accurately measured. Her work is covered in Chapter 3.

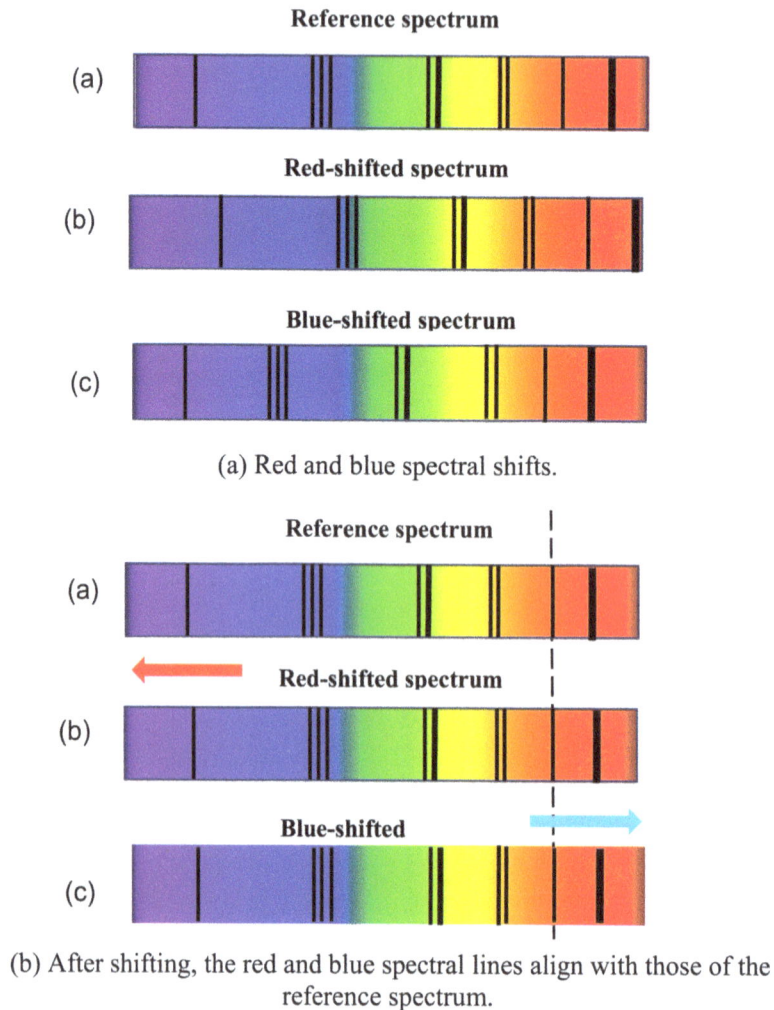

(a) Red and blue spectral shifts.

(b) After shifting, the red and blue spectral lines align with those of the reference spectrum.

Fig. 1.12 Redshift and blueshift in visible spectra.

Chapter 1. In the beginning…

Notes on Fig. 1.12

Fig. 1.12 (a) and (b) show three spectra each. The reference spectrum is for a nearby object, such as the Sun. The middle spectrum could be from a distant star or galaxy moving away from us. It shows a redshift, with the spectral lines shifted towards the red of the spectrum. The bottom spectrum is for a distant object moving towards us, e.g., Andromeda Galaxy. It shows a blueshift, with the spectral lines shifted to the blue end of the spectrum.

Fig. 1.12 (b) show how the spectra can line up when the redshifted spectra are shifted to the left, and the blueshifted spectra moved to the right.

Contrary to the general rule, the spectrum of Andromeda was found to be squashed rather than stretched. This is because, as we said above, it is a 'nearby' galaxy, part of the Local Group, and moving towards the Milky Way due to gravitational attraction.

We learnt earlier that the red colour light has a longer wavelength than the other colours. So, it appeared that for most objects the colours were being made redder the further the objects were from us. We call this the 'redshift'. An object's redshift is also used as a measure of distance.

Hubble had found an easy way to find out how far distant objects like galaxies were from us here on Earth. All he had to do was to see how much the light waves were stretched by examining the spectrum.

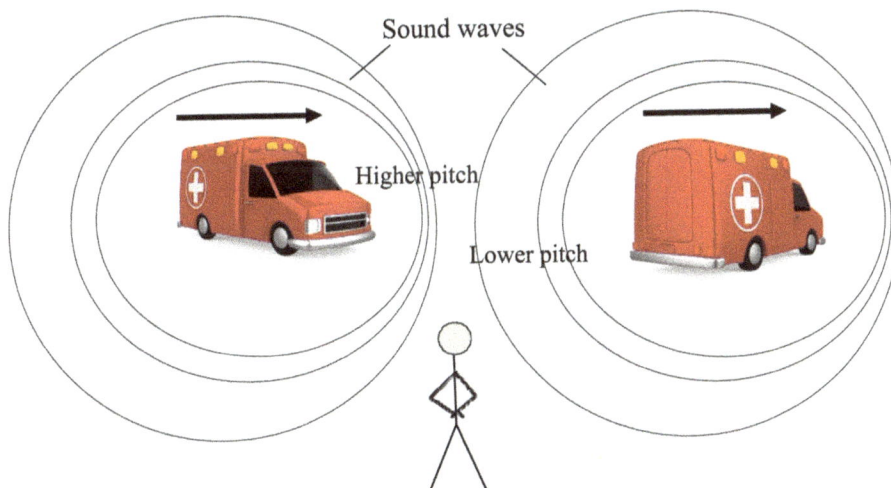

Fig. 1.13 The Doppler Effect.

You may have heard an ambulance, or a police car go by with its siren blaring. You would have noticed that the sound of the siren is different when the vehicle is coming towards you compared with the sound when it is going away from you. The sound is of a higher pitch (frequency) coming towards you and of a lower pitch when it is going away. The reason is that sound, like light, also travels in waves. These waves bunch together [Fig. 1.13] when the vehicle is moving towards you and get stretched out when it is moving away. This is called the Doppler effect (or the Doppler shift) after Christian Doppler (1803-1853), an Austrian physicist, who first explained the reason.

The same thing happens with light. Light waves get stretched if the source of the light is moving away from us and bunched up when the source is travelling towards us. Putting Hubble's redshift, and the Doppler effect together, the scientists came to an amazing conclusion. Because the frequency (pitch) of the light from the distant galaxies is lowered, it must mean that the galaxies are moving away from us. And since the redshift is bigger the further that the galaxy is from us, the further galaxies must be moving away faster than the nearer ones.

Hubble realised another amazing thing: *this can only happen if space is expanding*!!

Since the space is expanding and the speed of light is constant, all that can happen to light is that its waves get stretched, that is its wavelengths increase. The light therefore becomes redder, giving us the redshift.

Think about it. This means that we are not living in a static, steady Universe, but one that is expanding, getting bigger by the second.

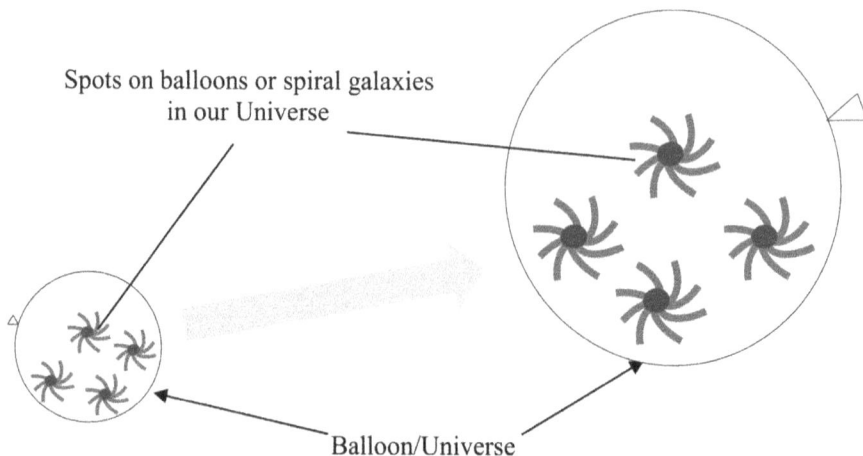

Spots on balloons or spiral galaxies in our Universe

Balloon/Universe

Fig. 1.14 Galaxies move apart as the Universe expands.
So, do spots on a balloon as the balloon expands.

A good analogy is to imagine that the Universe is like the surface of a balloon which is being blown bigger and bigger. You can do a simple experiment to see what happens. Take a balloon and mark some dots on it in ink. Now inflate the balloon. You will see the spots move further

Chapter 1. In the beginning…

away from each other the bigger the balloon gets, and the further that the spots are from each other, the more they move apart [Fig. 1.14].

This discovery was astonishing. Space is expanding, the Universe is getting bigger. But the idea that came next was truly astounding – if the Universe is expanding, it must have been smaller in the past. So, scientists started working backwards to see what happened. They came to a startling conclusion that the Universe at some time in the past would have been a single point. It must have started from nothing.

There must have been a Big Bang.

Moreover, knowing how fast the Universe is expanding, scientists could estimate when the Big Bang would have occurred, or, in other words, they could work out the age of the Universe. The original estimate continues to be refined as more accurate measurements are made. The current thinking is that the Universe came into existence in a Big Bang some 13.8 billion (that is 13,800,000,000) years ago.

We have seen that telescopes enable us to see the distant objects in space as they were in the past. Using the Hubble telescope, astronomers have seen galaxies which are more than 13 billion light years from us, the Universe as it was more than 13 billion years ago. Since the age of the Universe is believed to be about 13.8 billion years, we can see galaxies when the Universe was only around 800 million years old!

Before then there were very few stars at all. But the story from the Big Bang to the stars forming is for the later chapters.

Earlier, we spoke about the number of stars in the Universe, to show how big the Universe is. Now that we have discussed light, and its speed and how the space itself is expanding, I can introduce you to the final amazing fact for this chapter.

If the Universe is 13.8 billion years old the furthest object, we should expect to see would be 13.8 billion light years away. In other words, we would expect our observable Universe to be 13.8 billion light years in radius. However, the space itself has been expanding as the light has been travelling to us. In the 13.8 billion years, objects that were 13.8 billion light years away, when the light that has now reached us started, are by now be some 46 billion light years away. So, our *observable Universe* is 46 billion light years in radius, or 92 billion light years across [Fig. 1.15].

[Note: Space can expand at any speed. It is not matter and is not limited by the speed of light]

We have no way of knowing about what lies beyond this observable Universe, except to wait till that light has had time to reach us. We will be able to see additional light years, depending on the speed of expansion, in every direction, for each solar year that passes.

Light starts
towards us 13.8
billion years ago

A

Earth

Light from **A** reaches
us when **A** is 46
billion years away

A

Earth

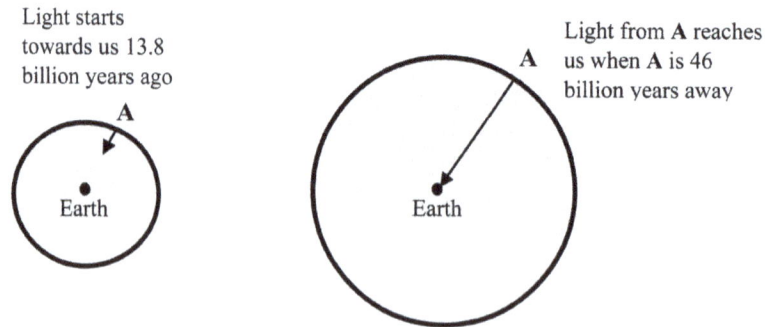

Fig. 1.15 The size of the observable Universe.

You should have realised by now that when we talk about the Universe, we are talking about the observable Universe, and not the *hidden* Universe beyond. The actual Universe may be far larger than the one we can observe, perhaps even infinite. We just do not know. But as time passes, we will see more of it while the expansion continues.

How big is the Universe?

The truth is, we do not know. Reasonable estimates have been developed for the Universe, the galaxies within it, and the stars within the galaxies, which is the best we can do. The figures we have worked to in this book are given below. Note that they are estimates and only refer to the observable Universe. How big is the total Universe is a mystery; it may be infinite.

The observable Universe is approximately 92 billion light years across. The number of galaxies in the Universe is about 200 billion but there may have been 10 times as many in the early days of the Universe. We must remember that the early galaxies were smaller. Over time, they merged into larger groupings. The number of stars per galaxy is 100-200 billion. Our galaxy, the Milky Way is 100-150,000 light years across, and it contains some 200 billion stars.

NASA's Hubble telescope was deployed during September 2003 to January 2004 to take images of deep space galaxies. The *Hubble Ultra Deep Field* image shows nearly 10,000 galaxies ranging from the time when the Universe was 800 million to 1 billion years of age, some 13 billion years ago. To get a sense of the grandeur of the galaxies, look at the *Hubble Deep Field* image full-sized at *https://www.spacetelescope.org/images/heic0611b/.

Remember each point of light there is not a star but a galaxy containing billions of stars.

---

* The availability of the site was checked in January 2021.

Chapter 1. In the beginning…

How did we get here?

We have taken a while to get here. But we have had to learn a lot of things along the way and answer many questions to answer the question 'how do we know there was a Big Bang?'.

We talked about the Universe being space filled with matter, and light in all its forms. We have seen that all stars, galaxies, gas and dust in space and all things on Earth including us, are made of matter. Matter itself is made of tiny atoms of different elements such as hydrogen, carbon, oxygen, iron, and gold. The atoms have a nucleus of protons and neutrons which have even tinier electrons going round them in a cloud; atoms combine to form molecules.

The Universe is also filled with electromagnetic radiation we call light. Some of this radiation we can see. When light and atoms collide, the light can get absorbed by the atom if it is the right frequency for the atom. This absorption is seen as black lines in the spectrum of the light. This tells us which elements the light beam met on its way to us and enables us to know what the stars and distant galaxies are made of.

Light travels at the same speed no matter whether it is coming towards us or moving away from us, independent of our own speed.

We use telescopes to see far into the distance. Telescopes allow us to see the objects as they were when the light left them in the past. Hubble found that the light from distant galaxies was 'redder' than light from nearer galaxies. This is the redshift. He concluded that the light waves coming from the distant objects were being stretched because the objects were moving away from us, and that the further the objects were the faster they were receding. This can only happen if space itself is expanding, and the Universe getting bigger. By working backwards, scientists saw that the Universe was smaller in the past and calculated that it started from a Big Bang about 13.8 billion years ago.

Finally, the light that left the early Universe has in fact taken 13.8 billion years to reach us. However, because the space itself is expanding, the objects we are seeing now are about 46 billion years away. This limits the size of the Universe we can observe today. With each year that passes this limit grows larger, of course, as the light from even further away reaches us from every direction. Space has no limit on its speed of expansion; it can expand faster than the speed of light.

Chapter 2

The next 400,000 years...

from the Big Bang to the first light

In this chapter we are going to see what happened immediately following the Big Bang, until we get to a Universe which we can begin to recognise with its stars and galaxies.

In the simplest terms, the huge amount of energy that the Big Bang created resulted in the Universe growing extremely large, exceedingly quickly. It is as if suddenly a vast amount of air was blown into our balloon Universe. The Universe started off extraordinarily hot but cooled as it expanded, and some of its energy changed into particles that went on to make up matter such as atoms.

But I want to give you a little bit more of the detail than this, to try and pass on the magic of what we think happened over a period of almost 400,000 years before the Universe blazed with its first light. Then I will tell you how we know what we know and, hopefully, clear up some of the comments that I made above, such as 'energy changed into particles'.

Since we are talking about the Universe, we are dealing with tremendously large numbers, but, because we are also talking of extremely short periods of time over which some of these changes happened, we are equally dealing with exceedingly small numbers. I will try and show you how we deal with such numbers in science and explain what happened after the Big Bang as simply as I can, taking it a step at a time. It should be interesting, and hopefully not too complicated to follow. If you are already familiar with these concepts feel free to skip the section. Or if you prefer, come back to it later.

The first trillionth trillionth trillionth of a second

At the instant of the Big Bang, the Universe was infinitesimally tiny, a point without any size, something we call a singularity. But even though it had no size, it did have an unbelievably large amount of energy. Now the more energy something has the hotter it is. So, the infant Universe was infinitely tiny, with an infinitely large amount of energy, at an infinitely high temperature.

When you have a large amount of energy in a tiny space, you get an explosion. In an explosion, such as is created by a bomb, the chemicals in the explosive are changed rapidly into a hot gas, releasing a large amount of energy in the form of heat, light and a shock wave which is due to the air surrounding the bomb suddenly being pushed away as the gas expands extremely rapidly. That is what makes an explosion so powerful. But as gas expands, the energy it has gets spread over a larger and larger volume; the result is that soon the temperature drops, and the power of the explosion fades away. Thus, while someone unfortunate to be near the explosion can be badly hurt or even killed, others far enough away can see the explosion without sustaining damage themselves. After the Big Bang though there was no place to stand and observe safely. The Universe itself was the explosion, being created as it expanded.

Even the biggest explosions that we humans can make, such as with nuclear bombs, are only an extremely poor comparison with what happened in the early Universe. It is difficult to imagine how powerful the Big Bang was.

When we talk of the Universe, one of the things that we must get our heads around is the scale of the things we are dealing with. Some things are tiny, while others are exceedingly large indeed. Scientists have developed ways of coping with this.

Dealing with extremely large and exceedingly small numbers

[You may skip this section and return to it later]

The Universe immediately after the Big Bang was extremely hot. Scientists estimate that its temperature was more than 10^{32} K. What does this mean? 10^{32} means 10 multiplied by itself 32 times. It is written as 1 followed by 32 noughts which is a hundred million trillion trillion. A hundred is written as 100, a million as 1,000,000, a billion which is a thousand million as 1,000,000,000, and a trillion which is a million, million is written 1,000,000,000,000. You can see that a hundred million trillion trillion is an unbelievably huge number. Rather than write this out in its long form, with its 32 noughts, as 100,000,000,000,000,000,000,000,000,000,000, the scientists use a short-form and write it as 10^{32} (we speak of it as 10 raised to the power 32). [Incidentally, if you tried to count to 1 trillion (10^{12}) at the rate of 1 number per second non-stop, it would take you about 32,000 years!]

Chapter 2. The next 400,000 years…

We deal with small numbers less than 1 in a similar fashion, but we 'raise' 10 to a negative number. When 1 is divided by 10, we write it as 10^{-1}, when we divide it by a second 10 (in other words when we divide 1 by a 100), we write it as 10^{-2}. So, for example, we write 10^{-36} seconds to signify 1 second divided 36 times by 10, which is an extremely brief period.

Let me try and show you how short 10^{-36} seconds of time really is: 1 second is the tick of an old-fashioned clock which you may have at home. A second is also about the time it takes you to say, 'one hundred and one'. In our decimal system, the next step down from a second is a tenth of a second which scientists write as 10^{-1} seconds and which can also be written as 0.1 seconds (there are of course ten of these in a second). Similarly, 10^{-2} second is a hundredth of a second or 0.01s, which is a decimal point with one zero before the 1. Thus, 10^{-36} seconds, which is a trillionth trillionth trillionth of a second, is 0.000000000000000000000000000000000001 seconds! That is, a decimal point with 35 zeros before the 1. In other words, by the time you have said 'one hundred and one' for the 1 second count, a trillion, trillion, trillion 10^{-36} seconds have passed! You can see how small the number is and why the scientists find it easier to write it as 10^{-36}, rather than in its longer form.

[End of optional section]

We saw above that temperatures measure the amount of energy something has. We also said that scientists estimate that the temperature of the Universe immediately after the Big Bang was more than 10^{32} K. We have seen above that 10^{32} is an exceedingly large number. But what is K?

You would have used a thermometer when you were ill to measure your body temperature, to see if you had a fever. Such a thermometer measures the temperature in Celsius which is written as C or sometimes as °C (called degrees C). [Some countries still use the older Fahrenheit temperature scale, but we will not go into that here]. For scientific work we tend to use other more accurate techniques than a home thermometer to measure a much wider range of temperatures, and we record the temperature in a different unit called Kelvin (K). Each degree C and degree K are of the same size, but their starting point is different. For comparison, water freezes at 0 C which is 273 K, and boils at 100 C which is 373 K. [0 K the starting point of the Kelvin scale has been chosen at the lowest temperature that is ever possible. It is called *absolute zero*. On the Celsius scale 0 K is -273 C. This is the temperature when all motion of atoms stops. Movement gives atoms their energy, and therefore when this stops there is zero energy, the atoms freeze, and it can get no colder.]

The Kelvin unit was named after the British physicist and engineer Lord Kelvin (1824-1907), who did pioneer work in defining the laws of physics which deal with energy and temperature.

Now, after our side trip into the world of numbers, let us go back to our story.

Immediately after the Big Bang we therefore had this tiny, hot Universe of seething energy. At that time, there was no matter in the Universe, it was just too hot for any particles of matter to exist. We have no idea what really existed then. But there may well have been extremely high energy particles (photons), colliding into each other in the tiny space.

Then something amazing happened.

Scientists believe that in the period between about 10^{-36} and 10^{-32} seconds after it came into existence, the Universe grew an unbelievable 10^{26} times (that is 1 followed by 26 noughts) to about the size of a grapefruit. You could have played football with the Universe then! This period of explosive growth is called *inflation*. We do not quite know why it happened, but the concept answers many crucial questions about the structure of the Universe. It was developed into a *theory of inflation* in the early 1980s. So far, the theory has stood up to all that has been thrown at it. In 2001, it earned Alan Guth (1947-) of the Massachusetts Institute of Technology (MIT), Andrei Linde (1948-) of Stanford University, and Paul Steinhardt (1952-) of Princeton University, the prestigious Dirac prize for its development.

According to the theory, the dramatic expansion smoothed out the Universe in every direction, so that all the energy was evenly distributed with only tiny irregularities left behind. Therefore, when we look at the Universe with a powerful telescope, we find that it appears the same wherever we look, with stars, galaxies and groups of galaxies spread out in the same way in each direction. However, the tiny irregularities that were left behind proved to be important. As we shall see in due course, without them, there could have been no galaxies, no stars, and no planets, and no us.

As the Universe expanded due to inflation, it also cooled, just like the gas in an ordinary explosion would cool. The scale is different, of course. During inflation, the Universe's temperature dropped from 10^{32} K to about 10^{28} K. It is about ten thousand times cooler, but still exceedingly hot. This is cool enough though for the first particles of matter to condense out of the energy.

Among the particles that were created then were quarks, which as we already know from the last chapter are the fundamental building blocks of matter, and electrons. Another important thing that was produced at this stage was the mysterious dark matter. You will shortly see what a crucial role dark matter plays in putting the jigsaw of the Universe together.

I said that particles of matter 'condense out of energy'. At the start of the chapter, I also talked about energy changing into particles. What does this mean? How can matter be produced out of energy?

Chapter 2. The next 400,000 years…

There is an extremely important concept in science that was discovered by Einstein (who else). Einstein showed that energy and matter are just two different forms of the same thing and can be converted from one into the other. He came up with what is without doubt the most famous equation in science, one that possibly you have come across already. This equation, almost the only one you will see in the book, is:

$$E = m\ c^2$$

This says that E (energy) is equal to m (mass, which you should remember from Chapter 1 is the measure of how much matter there is in a body) multiplied by c (speed of light) squared. Squared means a number multiplied by itself; thus 2 squared (which can also be written as 2^2) is 2 x 2 or 4. In other words, the equation states that E is m x c x c. We saw in the previous chapter that light travels at a speed of 300,000,000 metres per second. Squaring 300,000,000 gives us 300,000,000 x 300,000,000 or 90,000,000,000,000,000. This is an exceptionally large number indeed. (Note that to multiply these numbers we just multiplied 3 by 3 and added up all the zeroes to give us the number 9 followed by 16 zeros. In other words, $c^2 = 3$ x 10^8 x 3 x $10^8 = 9$ x 10^{16}).

Thus, if we change even a tiny amount of matter (m) into energy we get a lot of energy (E) created because of the large value for c^2, resulting in a remarkably big bang indeed. This is how atom bombs and nuclear power stations work. A tiny amount of matter of a radio-active element, such as uranium or plutonium, is converted into a large amount of energy which in the case of a bomb comes out with a bang, but which in a nuclear power station is controlled to give us energy in the form of electricity for our homes.

The opposite can also happen. The same equation, looked at in the opposite direction, shows that a large amount of energy can convert into a small amount of matter. Energy and matter are truly interchangeable.

So as the early Universe expanded, its energy (and temperature) dropped, and the energy that was lost was converted into an equivalent amount of matter in the form of tiny particles such as quarks and electrons. Matter was thus created, not out of nothing, but out of energy.

In nature, particles do not exist in isolation. They interact with each other. But how? We shall now look at the four fundamental forces of nature that cause the interaction between particles of matter, and how they came into being.

The fundamental forces of nature

[You may skip this section and return to it later]

There are four fundamental forces of nature. One of these forces that we have already talked about is the force of gravity, or the **gravitational force**, between masses; this is the force by which masses attract each other. We saw in Chapter 1 that the larger the mass of the objects and the nearer that they are to each other, the bigger is the gravitational force of attraction between them. Gravity always *attracts* objects to each other and can act over infinite distances. But by itself it is the weakest force of the four fundamental forces in our Universe.

As its name implies, the **strong force** is the strongest of the four fundamental forces. It is strong enough to hold the nucleus of an atom together. It acts over the tiny distances (10^{-15} m, where *m* stands for metre) found in an atom. This is the force we must overcome when, for example, in a nuclear reactor we split an atom to produce energy.

The **weak force** acts over even shorter distances (10^{-17} m) than the strong force. It is responsible for some of the atomic level interactions, such as those that change a proton to a neutron. As its name implies, the weak force is not as strong as the strong force being about a million times weaker.

The **electromagnetic force** causes the interaction between electrically charged particles. This is the force that is responsible for holding atoms and molecules together and keeping electrons in orbit around the atomic nuclei ('nuclei' is the plural of the word 'nucleus'). It behaves in a similar manner to gravity in that it can act over an infinite range, and, again as with gravity, the larger the charges are on the objects or the nearer they are to each other, the bigger is the electromagnetic force between them. However, while the force of gravity always attracts, the electromagnetic force attracts when the charges are dissimilar (one positive and the other negative) but repels similar charges (where both are positive or both are negative). We shall further discuss atoms and their nuclei later in this chapter.

At the time of the Big Bang all these four forces were combined into one. As the Universe expanded and cooled, and its energy dropped, the four forces separated out, one after the other. The first force that separated out was gravity, even before inflation started. The strong force split away just after inflation and finally the electromagnetic and weak forces parted company. 10^{-12} seconds had passed after the Big Bang by the time the forces spilt into four, and the temperature was now down to 'only' about 10^{16} K.

[End of optional section]

Matter and antimatter

'Matter' comes in two forms in nature: matter and antimatter. For every matter particle, there is a corresponding antimatter particle – there are quarks and anti-quarks, electrons, and anti-electrons (called positrons), and so on.

Antimatter is formed by its own set of antimatter particles: while electrons are formed of quarks, positrons are formed of anti-quarks, etc. Antimatter is the same as matter except that it has the property that when it meets its equivalent matter particle, they annihilate each other, both disappearing, their masses having been converted into pure energy (in accordance with

Chapter 2. The next 400,000 years…

Einstein's equation that we saw earlier). But we live today in a Universe almost entirely made up of matter. What happened to the antimatter that was created?

The Big Bang created both types of particles, matter, and antimatter, in *almost* equal numbers. Luckily for us, it appears that slightly more matter was created than antimatter, so that after all the collisions between the matter and antimatter particles had occurred, enough matter was left over to make the stars and galaxies, and us. That is why there is hardly any antimatter in the Universe compared with matter. A question that scientists are still trying to answer is exactly why more matter was created than antimatter.

Antimatter may be difficult to imagine but it really does exist (though if you ever come across your antimatter twin you will need to be careful that you do not touch each other, else *boom…*). Indeed, scientists can produce it in their particle accelerators. The problem is: how are we to store antimatter since it must not touch its matter particle twin, else both will be destroyed. In fact, there is a way to store these exotic particles. They can be kept in the middle of a high-vacuum container, held there by strong magnetic fields so they do not encounter matter.

But with our current technology, they can be stored only for exceedingly short periods. It would be useful if we could figure out how to make antimatter cheaply and to store it conveniently for long periods because then we could build the matter-antimatter engines for powering our spaceships, as happens in so many science-fiction stories.

Atoms and their nuclei

[You may skip this section and return to it later]

You should remember from Chapter 1 that an atom consists of a nucleus at its centre, and electrons whizzing around this as if in orbit. The nucleus is composed of protons and neutrons, which we came across in Chapter 1, and which are themselves made of the even tinier quarks that we also met earlier. The first nuclei made in the newly created Universe were those of hydrogen and helium. The nucleus of hydrogen is simply a proton. The helium nucleus is made up of two protons and two neutrons [Fig. 2.1].

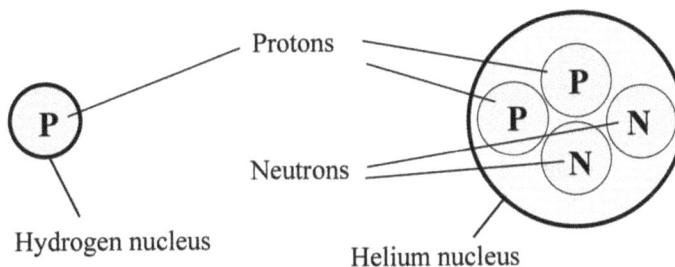

Fig. 2.1 Hydrogen and helium nuclei.

31

You may be familiar with the poles of a magnet. Every magnet has two ends, usually at the ends of a bar of iron or at the ends of a horseshoe shaped piece of iron. Each end of a magnet has a differently charged 'pole', one will be positive (+) and the other negative (–). If you bring the ends of two bar magnets near to each other, they will either attract each other and stick together if one pole is positive and the other negative, or if both poles are positive or both negative they will repel each other, and it will be difficult to get them even to touch. The way to remember this is to remember that unlike poles attract while like poles repel.

A similar thing happens with electrical charges. Unlike charges attract while like charges repel. The forces acting on the poles and the charges are due to the electromagnetic force we met earlier.

Each proton has a single positive electric charge, and each electron has a single negative charge. The neutrons, as the name implies, are chargeless (neutral).

A hydrogen nucleus has a +1 charge because of its single proton. The helium nucleus has two protons and therefore has a +2 charge. An electron on the other hand has a negative (– 1) charge and is attracted to a positive nucleus. A hydrogen nucleus will therefore attract one electron before its charge is neutralised. Similarly, a helium nucleus will attract two electrons to neutralise its charge. When the electrons team up with a nucleus with an equivalent positive charge and start orbiting around it, together they form a stable atom with zero-charge [Fig. 2.2].

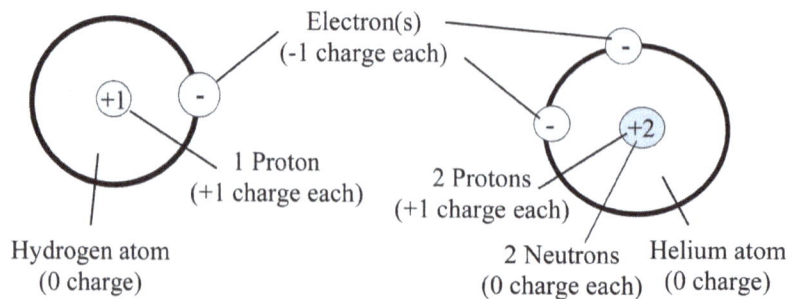

Fig. 2.2 Charges on hydrogen and helium atoms.

Thus, a neutral hydrogen atom has one proton in its nucleus and one electron in orbit, while a neutral atom of helium will have two electrons in orbit around its nucleus containing two protons and two neutrons. The same rule applies to all other elements that we know in the Universe – the number of electrons spinning around a nucleus of a stable, neutral atom is equal to the number of protons in its nucleus.

Chapter 2. The next 400,000 years…

There. You have just learnt an extremely important item of science about atoms which make up all elements in our Universe. I hope that you did not find it too complex to follow.

Protons and neutrons are similar. They weigh almost the same (the neutron is about 0.1% heavier) but they have one important difference. We have seen that while the proton has a positive (+ 1) charge, the neutron has no charge.

Protons are 1836 times heavier than electrons. Most of an atom is empty space: if you had an atom the size of St. Paul's Cathedral, the proton will be an orange at the centre, and the electron a fly buzzing near the roof.

[End of optional section]

The Universe continues to evolve…

Following inflation, we saw that the Universe had its first particles (and any remaining anti-particles) of quarks and electrons, plus the mysterious dark matter. The Universe continued to expand (but at a much slower rate than during inflation) and to cool as it did.

The fundamental particles such as quarks had been milling around since being created. The quarks cannot exist in isolation and soon started to combine when the temperature was down to about 10^{12} K (a few trillion degrees K) to produce the first protons and neutrons.

Particles continued to collide and react with one another, till between a minute and an hour after the Big Bang, when the temperature was down to a few 10^8 K (hundred million degrees K) the nuclei of the first complex atoms started to form by the combinations of the protons and neutrons.

Remember that the Big Bang produced not only matter particles, but also particles of antimatter. We also know that when matter and antimatter particles of the same type meet, they annihilate each other. Because of these collisions most of the electrons and antielectrons (also known as positrons) disappeared, releasing energy. However, there being slightly more matter than antimatter particles, some (but still an enormous number) of the electrons remained. These remaining electrons were still too hot then to combine with the nuclei to form atoms and were wandering free. Importantly, they were meeting and colliding, as in a fun-fair dodgem game, with the many more photons (particles of light) which were also filling the Universe.

These collisions kept the light photons bottled up; light was stopped from escaping. The Universe was growing as it expanded, however it was still dark, as if in a fog. But this was about to change. Dramatically.

Let there be light!

By 380,000 to 400,000 years after the Big Bang the temperature of the Universe had dropped to between 4500 K and 3000 K, a long way down from the 10^{32} K we met earlier. As the Universe cooled down, more and more of the wandering free electrons were captured by the nuclei of atoms that were missing their full complement of electrons (you will recall that each hydrogen atom can have one electron, and each helium atom two electrons). This event when the electrons combined with the nuclei is called *recombination*. Slowly the free electrons diminished in number, as more and more combined with atoms which were lacking in electrons. (Note: atoms which are missing some of their electrons are called *ions*; a gas of ions is called a *plasma*).

As the number of wandering electrons dropped, so the number of collisions between electrons and light photons decreased, and more and more photons could escape the fog. Slowly the fog lifted till about 380,000 to 400,000 years after the Big Bang, virtually all the photons had escaped, and the Universe blazed in its full glory. But, if we humans could view the Universe at this time, we would not see this light.

When I use the word light here, I of course mean electromagnetic radiation. You will recall from Chapter 1 that visible light (light to which the human retina is sensitive) is only a small part of the radiation spectrum. The frequency of the radiation in the rest of the spectrum is either too low or too high to be visible to our naked eyes. The light that now escaped had an extremely high energy since its photons had been produced when the Universe was exceptionally hot and energetic. If we could examine it then we would find that it was made up of *gamma rays*, which are the highest energy radiation, with the highest frequency and shortest wavelength.

The radiation left over after the *recombination event* still exists today and is called the cosmic microwave background (CMB) radiation. It fills our Universe and reaches us from every direction in space. There is an interesting story behind why it is called cosmic microwave background, and how it was discovered.

The cosmic microwave background (CMB) radiation

The CMB radiation had been predicted way back in 1948 by several physicists, including George Gamow (1904-1968) - a Russian scientist who later became an American citizen - by examining the physics of the Big Bang. But this prediction was subsequently almost forgotten till a lucky event occurred in 1965.

In the early 1960s, two young US scientists working at the Bell Telephone Laboratories in New Jersey, USA, Arno Allan Penzias (1933-) and Robert Woodrow Wilson (1936-), built an extremely sensitive instrument [Fig. 2.3] for some experiments in radio astronomy. This was a 6m (20 ft) horn (rather like the ear-horns that were used in the olden days as hearing aids, or like a Swiss cow-herd's horn). When Penzias and Wilson started using this instrument, they found that they were receiving a "hum" from all over the sky, wherever they pointed their instrument, day, or night. The hum had a wavelength of 7.35 cm.

They were convinced that there was a problem with their instrument and started checking it out thoroughly. They cooled the equipment down to -269 C (or only 4 K) using liquid helium. They even thought that the problem was because of some pigeons that were nesting in the horn, due to their poo coating the inside and generating the heat that was causing the wrong readings! They spent a long time scrubbing out inside of the horn and removing the pigeons, but all to no effect.

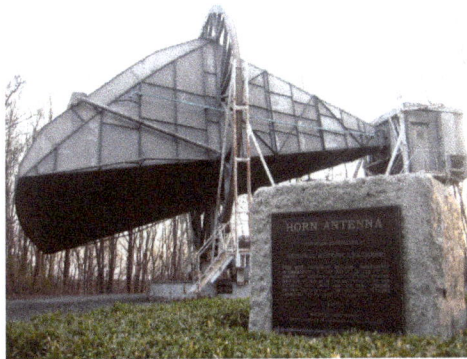

Fig. 2.3 The horn antenna 'telescope' used by Penzias and Wilson.

They then started asking around their scientific community to see if anyone may have an explanation.

It so happened that at the same time astrophysicists from Princeton University, only a few miles from the Penzias and Wilson experiment site, were preparing their own experiment to look for the microwave radiation in space. When they looked at the "errors" that had been found by Penzias and Wilson, they realised that the hum was the CMB; Penzias and Wilson had accidentally found the microwave radiation before them. The two sides published their results together, but in 1978 only Penzias and Wilson were awarded the Nobel Prize for their discovery. Sometimes that is the way the cookie crumbles in science. Sometimes life really is not fair.

But what was special about the radiation that was exciting so many scientists? The answer lay in the fact that the radiation was all around us and was at microwave wavelength (7.35 cm) which would be produced by atomic particles at a temperature of 3.5 K. Keep these figures in mind.

The scientists realised that if the Big Bang had really happened, and if their theories were correct about the radiation escaping some 380,000 years after the Big Bang, then this radiation must still be around. And since the event happened all over the Universe, we should see it wherever we looked. This was what was happening with the newly discovered radiation. However, the temperature of the radiation was 3.5 K, which meant that the radiation had low energy, and therefore a low frequency (and thus a large wavelength). Such a radiation would exist in the microwave part of the electromagnetic spectrum. But we had also said that the radiation that escaped from the dark Universe was of extremely high energy, and therefore of an extremely high frequency and short wavelength. The ancient radiation released when recombination occurred was at the gamma ray end of the radiation spectrum, while the radiation that was found was at the completely opposite microwave end. How could these two be the same thing?

This can be confusing until you realise that between the gamma rays escaping and the microwaves being found something had changed. Time had passed. In fact, over 13 billion years had passed. Remember that the Big Bang resulted in the expansion of the Universe. There is no reason to think that the expansion stopped. In fact, as we saw in Chapter 1, Hubble had shown that the Universe was still expanding, and we had deduced that the Big Bang happened by working backwards in time till we came to the single instant of the Big Bang that happened 13.8 billion years ago.

The Universe was expanding and cooling as time went by. When I say the Universe was expanding, I mean that space itself was expanding. And since the speed of light was remaining constant, the expansion of space was stretching the very fabric of light, the length of its waves. Therefore, as the time elapsed, the light waves got longer and longer, and their frequency dropped lower and lower. Over the 13.8 billion years this stretched the light of gamma ray wavelength, which is about 10^{-10} cm, to the microwave wavelength of 7.35 cm (that is the light waves were stretched about a 100 billion times). The temperature over this period dropped to just 3.5 K. This analysis matched the radiation being seen.

The discovery of the CMB enabled the scientists to do one other important thing. At the time it was discovered, there were two competing theories about the Universe. One was the Big Bang theory about which we have been talking. This said that the Universe came into being in a single cataclysmic event. The other was the Steady State theory (championed by the astronomer Fred Hoyle in the UK, amongst others), which said that the Universe had always existed in the same form as it is today, and that new matter was being created to fill the 'empty' space produced as the Universe expanded. The CMB showed that the Big Bang theory was correct since the Steady State theory could not explain the CMB.

The CMB is the picture of the Universe when it was 380,000 years old. It maps the irregularities left in the cosmos after the *Inflation*-caused expansion of the Universe. It must then surely have a lot to tell us about the period when atoms had formed but before there were stars or galaxies. It must tell us what the Universe looked like at that time, and we should be able to map this against what we have in the Universe today.

Indeed, this is exactly what the study of the CMB has done. Let me now tell you about this. I will also tell you how you may see (and hear) the CMB for yourself.

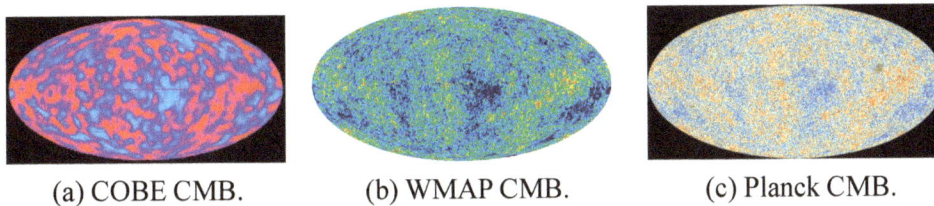

| (a) COBE CMB. | (b) WMAP CMB. | (c) Planck CMB. |

Fig. 2.4 Images showing the CMB temperature differences across the sky.

The CMB is so important that three telescopes were sent up successively into space with the express objective of studying the radiation. First, a telescope called COBE (Cosmic Background Explorer) was sent up in 1989 by NASA. It imaged the CMB and produced a map of the temperature differences over the whole of the sky, looking in every direction. This showed a speckled image [Fig. 2.4 (a)]. The image look starks with different pink and blue colours, but the colours are false and have been chosen to highlight the extremely tiny differences, or ripples, in the temperatures which were found. These variations are only about one part in 100,000. To give you an idea as to how smooth it is, consider an analogy someone has worked out which says that if the surface of the Earth were smooth to 1 part in 100,000, it would be smoother than the smoothest billiard ball and Mount Everest would be just 100 m high.

The WMAP telescope sent up in 2001 was another NASA project. It had a complicated name, the Wilkinson Microwave Anisotropy Probe. It did much the same job as COBE but at a much-improved sensitivity. The word *anisotropy* in its name means that measurements were of an item (temperature in this case) which has different values in different directions in space. It was named after David T. Wilkinson (1935-2002) an eminent US scientist who specialised in the study of the CMB. As you can see from Fig. 2.4 (b), the images this telescope obtained were much finer than the coarse COBE pictures.

Then in 2009, the European Space Agency (ESA) sent up the Planck telescope. This was named after Max Planck (1858-1947) a renowned German physicist who won the Nobel Prize for Physics in 1918. This telescope improved the results yet further and mapped the sky to an unprecedented sensitivity [Fig. 2.4 (c)]. It refined the results for the density of ordinary matter and dark matter in the Universe, as well as the age of the Universe.

Though the variations identified by the telescopes were extremely small indeed, matter slowly drifted due to gravity towards these sites of high density. It is thought that the first to fall into these high-density locations was the more extensive dark matter. Slowly, the mass of the collecting matter grew at these places, and as a result, the gravity increased, attracting even more matter. A snowball effect took hold and more and more matter collected as time passed.

Together, the CMB telescopes achieved many important results. They refined the age of the Universe to 13.77 billion years and showed that the Universe was not perfectly smooth but was anisotropic, with extremely slight differences in temperature in different directions. They proved that the matter made of atoms, with which we are familiar, forms less than 5% of the Universe, dark matter, about which we know virtually nothing, forms 24%, and dark energy, about which we are even more ignorant, forms the rest, slightly over 71%. One idea about dark energy is that it is the energy found in a vacuum and is responsible for pushing the fabric of space apart to expand the Universe. This is analogous to the air that is pumped in to expand the balloon Universe we talked about in Chapter 1 [see Fig. 1.14].

An interesting point is that the tiny temperature variations that the telescopes measured, equate to the tiny differences in the density of matter that was spread over the Universe. In other words, the CMB study showed that despite *inflation* matter was not evenly distributed in the Universe but showed minute variations with some locations being denser (having more normal and dark matter) with a greater gravitational attraction than others. These variations were crucial to the growth of stars and galaxies since they were regions in space where matter could collect in clumps.

Indeed, today we can map our existing distribution of galaxies to the tiny variations we see in the CMB temperatures.

In the next chapter we will see what eventually resulted from this gathering matter.

We have made many claims in this chapter: we said that at 10^{28} K quarks and antimatter were created; at 10^{16} K electromagnetic and weak forces spilt apart; at 10^{12} K the first protons and neutrons appeared; at 10^8 K atoms formed by protons and neutrons coming together; and between 4500 K and 3000 K the electrons were captured by the proton and neutron nuclei and atoms formed.

But how do we know all this?

How do we know?

There are two approaches we can adopt in answering the claims. In both, we work backward in time and consider how much energy is required to break something apart. We can then infer that if the energy is less than this, the particles will stay together without being pulled apart.

One approach is theoretical. We can work out how much energy is required for the different interactions to take place. For example, we can calculate the energy required to extract an electron from the pull of the nucleus of an atom. If we know this, then we know that if the energy is less than this, the electrons will stay put with the atoms. Similarly, we can calculate the strong force that is holding a proton and neutron together in an atom nucleus. Then we know that if we can provide more energy than this the protons and neutrons will be separated, or looking at it the other way, if there is less energy than this, the two particles will come together if they are near enough to be captured by the strong force and they will then stay 'glued' together until a force stronger than the strong force pulls them apart.

The second approach is experimental. We conduct experiments to see what are the energies required to split existing pairs. How much energy is needed to strip an electron from an atom; how much is needed to pull apart a proton and neutron; etc. We can provide this energy in many ways. If the energy is low, we can heat the substance, or perhaps use electrical or magnetic means to pull particles apart by means of their charge, for example electrons from atoms. But for anything significant, we use the remarkable machines that are called particle colliders.

The Large Hadron Collider (LHC) was in the news in 2012-13 when it was used to prove the existence of a special particle called the Higgs boson which gives other particles their mass. [This particle had been predicted by Prof Peter Higgs (1929-) of the University of Edinburgh in 1964. He was present at the announcement of the discovery by the LHC in 2012. Prof Higgs was awarded the Nobel Prize in Physics in 2013.]

The Large Hadron Collider (LHC), which cost £2.6 billion to build, is an amazing device. It is a particle accelerator, a machine in which two beams of atomic particles are accelerated to extreme speeds in opposite directions in circular pipes before being made to collide. The LHC was proposed in the 1980s but engineering work started only in 1994, and it became operational in 2010. It is the most powerful particle accelerator built up until the time of writing this book. Its power will be exceeded by the next generation Future Circular Collider (FCC) already being planned.

The LHC is in the form of a circular tunnel 27 km (more than 16 miles) in circumference situated near Geneva, straddling the Franco-Swiss border. (NB LHC's successor, FCC, is planned to be 100 km (over 60 miles) around). In the LHC, the beams of particles, usually protons, are accelerated by powerful magnets till they reach nearly the speed of light and achieve tremendous energies. Then they are made to collide. The particles that are produced in the debris are tracked and measured by huge, complex detectors to identify what they are. The energies at which they were produced are also measured by the machine.

A lot more work is planned for the Large Hadron Collider. Ultimately using the LHC, or FCC its successor, we should be able to achieve the extremely high energy conditions that existed nearer to the Big Bang itself, enabling us to answer many of the other questions that today are mere speculations.

Seeing (and listening to) the CMB radiation

I said during the discussion of the CMB that I will tell you how you can see the cosmic background radiation for yourself. Here is how.

It so happens that a TV screen can show us the CMB right in our living rooms. However, unfortunately, this is the case only if you can get hold of an old analogue TV. Some of us still use a rod-type of aerial for our TVs rather than satellite dishes. Such an aerial (or antenna as it is also called), is sensitive to microwaves and will capture them and send them down to your TV sets.

When you are watching a programme on TV, the set is tuned to the signal from the TV station and all other frequencies of the signal are separated out. However, if your TV is detuned, in other words, it is not tuned into a particular station, you get a lot of static or noise (also called 'snow') that you see as mush on the screen. A small part (1% or so) of this noise is in fact the CMB.

Unfortunately (for our experiment), we have now gone digital in most of the world with our TV transmissions. The digital technology is based on the binary (1, 0) system used in computers. This makes it impossible, to see this noise. So, it may not be possible to see the CMB on your own TV. But it is worth a try to see if there is an old TV still hanging about your home and plug in the aerial.

You can also hear the CMB. It is a part (about 1%) of the hiss (also called "white noise") you get if you tune your radio to frequencies between stations.

Alternatively, and rather more easily, you can search on the internet for "cosmic microwave background radiation sound" and listen to it on YouTube! You should also be able to see and hear the hiss of the untuned analogue TV on that channel.

Chapter 3

Twinkle, twinkle little star…

how I wonder what you are

We left the last chapter with light blazing freely across the Universe, and with dark matter and the atoms of matter, chiefly hydrogen, and helium, slowly falling into the denser areas in space. What happens then is the one of most exciting parts of the story of our Universe.

This chapter starts with the matter slowly accumulating in the ripples in space left behind after inflation. It will end with the Universe formed pretty much as we know it now, full of galaxies, stars, and planets. We will see what a star is, how it forms, lives, and dies. And we will understand how important each star generation has been in producing a Universe that could support life. We will see the many different types of stars there are, from the smallest to the gargantuan. Along the way we will come across white and brown dwarfs, red and blue giants and supergiants. We will learn about the enormous energy in the stars, and see how some become the brightest objects known, shining with a light that can be seen across the Universe.

Let us carry on our journey across time and space.

The first stars and galaxies are born

Once the electrons had combined with the atomic nuclei to form atoms and the light photons could escape and speed through space, the Universe consisted of light, dark matter, and normal matter - chiefly in the form of gas which was virtually all (about 90%) hydrogen and a little (under 10%) helium, with traces of deuterium and lithium. The period of inflation had left space in an exceptionally smooth state apart from ripples which were denser regions. The dark matter slowly fell into these denser regions due to gravity. The gas followed the dark matter. This accumulation speeded up as more and more material collected, and its gravitational attraction increased on the other nearby gas and dust particles.

Enormous clouds slowly accumulated in the cold space. The gases in each of these clouds continued to fall inwards to its centre pulled by the gravitational attraction of the zillions of gas molecules.

[Note: We saw in Chapter 1 that molecules of a substance are made of atoms of various elements that combine to form the basic building block of that substance. For example, each hydrogen molecule consists of two hydrogen atoms (written as H_2), a water molecule consists of two atoms of hydrogen and one of oxygen (written as H_2O) [Fig. 3.1].]

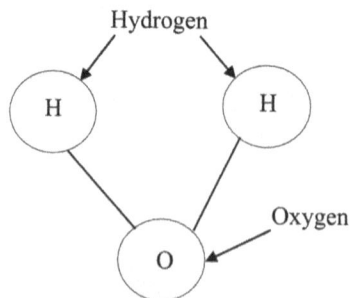

Fig. 3.1 A water molecule.

The pressures at the centres of the gas clouds built up as the huge volumes of gas were compressed due to gravity, and the temperature rose as the collisions between the gas molecules increased. You may have found this happen for yourself if you ever used a bicycle pump: as you fill up a tyre, the pump gets harder to push because of all the air already in the tyre pushing back, and the air near the nozzle of the pump gets hotter; or even cupped your palms together and blew into them and felt your palms get warmer the harder you blew. Similarly, as gravity pulled more and more of the gas into the centre of the cloud, the pressure there became higher and the gas there got hotter.

As the cloud collapsed, it slowly started to rotate and flatten into a saucer shape. Such rotation happens naturally when a volume of fluid (gas or liquid) falls towards its centre. A whirlpool forming in a river, or water spinning as it flows down a plug hole in a sink are examples of this rotation. The spinning, falling cloud got hotter and hotter at its centre. However, if you were outside the cloud, there was so much gas in the way that it would have made it impossible to see anything at the centre. But hidden away, some interesting things were brewing.

The pressure at the centre rose till it reached a critical point and the gas was hot enough for nuclear reactions to start. When that happened, the compressed mass of gas at the centre of the cloud suddenly ignited, began to shine, and give out enormous amounts of energy. A new star was born.

Over a period of millions of years, the Universe was teeming with stars. Because there was so much gas around when the first stars were born, these in general were much bigger than most of the stars today. Stars of 50 to 100 solar masses (50 to 100 times as massive as our Sun), and more, were common.

The newly formed stars gathered in clusters due to the gravitational attraction between them. The clusters collected in larger and larger groups till they contained millions of stars. The groups then came together and started to form even larger collections of billions of stars that today we call galaxies. The movement of stars in the galaxies formed a common pattern and the galaxies started rotating, much like the gas from which the stars had earlier formed.

Typically, the galaxies formed a spinning disk-like structure with a bulge in the centre containing an enormous number of stars. The gas clouds in some galaxies were in the 'arms' around the central disks. Such galaxies are called spiral galaxies for reasons that are obvious from their images. Some galaxies were round and are called globular galaxies. Others were elliptical or even irregular. Fig. 3.2 shows the main types of galaxies – spiral, elliptical, and irregular – but many other shapes and variations are to be found.

The Universe was still expanding, though at a *much* slower rate than occurred during the period of inflation. The galaxies that formed were being pushed apart by the expansion. But there were many that were near enough to each other to be bound together in groups by their own gravities. These galaxy groups therefore overcame the pushing apart that was being imposed by the expanding Universe. Our galaxy, the Milky Way, belongs to a group called, logically enough, the Local Group. There was a lot of gas, together with dust, in amongst the stars. New stars were still forming in any collapsing regions within such gas clouds. But stars were also dying.

(a) Barred spiral galaxy
NGC 1365.

(b) Elliptical galaxy
NGC 1316. (Credit: ESO)

(c) Irregular galaxy
IC 559.

Fig. 3.2 Main types of galaxies.

It so happens that the larger the star, the shorter is the time that it lives. While stars the size of our Sun can live for many billions of years, stars of 50 to 100 solar masses may last only a few million years. When these giants die, they do so spectacularly in huge explosions called supernovae: the star explodes and spreads its material all over nearby space. Gas clouds which are in the vicinity of a supernova can be given a shove by the shock wave from the explosion and start to contract so beginning the next generation of star building. Importantly (and we shall see later why this is so important), the gas that forms the next generation of stars contains not only the primordial hydrogen and helium gases, but also material that originated in the nearby exploding star.

The galaxies that formed contained many types and sizes of stars. They ranged from tiny suns barely able to shine, to the giants we met above. The stars, when they died, left behind remnants: some (white dwarfs) were about the size of the Earth but millions of times more massive; others (neutron stars) were even smaller, about the size of a town, but many times yet more massive; a few (black holes) were extreme. Shortly we shall learn more about these strange objects.

Some stars can be born in clusters near each other out of one huge gas cloud. We can see many examples of stars clusters in the night sky. An easy one to spot is called the Pleiades, also known as the seven sisters, shown in [Fig. 3.3], which is an open cluster of about 3000 stars, 400 light years from Earth in the constellation of Taurus the bull.

Some stars may be born out of the same gas cloud in binary systems, near enough to end up orbiting each other. Some can even be in triple star systems, or more. Imagine, a sky with many suns. Perhaps there is such a system out there in which at least a couple of suns are always in the sky of an orbiting planet, so that there is never night on this planet. No one there would have seen a star (apart from their own suns, of course).

Chapter 3. Twinkle, twinkle little star…

Fig. 3.3 The Pleiades cluster.
(Also known as the seven sisters.)

Isaac Asimov (1920-1992), a famous writer of science fiction stories, wrote a book about a planet in a six-sun system which results in a world that has night only once every 2049 years. The book tells about what happens to its inhabitants when darkness falls due to the rare eclipse of its sun while the other suns are hidden from that part of the planet. The book is called "Nightfall" and is well worth reading. Stop press: In January 2021 NASA announced that its Transiting Exoplanet Survey Satellite (TESS) had found a 6-star system about 2000 light years from Earth. They called it a sextuply-eclipsing sextuple star system. It is the fourth such system known.

Many of the stars when they form, end up surrounded by a disk of gas and dust from the cloud which gave birth to them. Soon after birth, the star creates a 'wind' of charged particles blowing away from it. This wind eventually blows away the gas cloud that created the star, leaving behind the sun shining in its glory and, perhaps, surrounded by a system of rocky and gas planets which had formed within the disk of dust and gas. We will talk more about the planets and how they form in Chapter 5 which is devoted to our own Solar System. Here we are concerned about the stars. But what is a star? What makes it shine? How do they evolve? How do they die?

That is the next part of our continuing story.

What is a star?

We have all seen stars. Even in broad daylight. Our nearest, best-known star is, of course, our very own Sun. During the day it shines so brightly that virtually everything in the sky is drowned out by its light. The stars we see at night appear as points of light, some brighter than the others, and some (if you look carefully) a different colour. In fact, all stars are basically the same: balls of glowing, bubbling gas; their colour determined by how hot their surface is.

The brightness of stars as seen by us depends not only on how bright they are, but also on how far they are. The nearer a star is to us, the brighter it appears, just as a light from a torch carried by someone walking towards you gets brighter the nearer the person gets to you. When we measure the distance of the stars from us (I will tell you later how we can do this), we find that many of the stars are so far from us that their light can take thousands of years or more to reach our naked eyes. Our telescopes allow us to see stars much, much further away.

But not all the shining objects in the night sky are stars. Some of the other objects we see that shine by reflecting the light they receive from the Sun. One of these is our Moon. Other objects we see in the night sky that are not stars include the planets and, if we are lucky, comets. We shall learn more about these objects when we talk about our Solar System in Chapter 5. Here we are interested in the stars.

Stars are objects in space that shine by their own light, and not by reflected light like the Moon or the planets. Stars are also large, much larger than the planets and the other smaller bodies one finds in space.

There are countless stars in the Universe. They exist in huge groups called galaxies which can contain hundreds of billions of stars. And there are hundreds of billions of galaxies in our observable Universe. Our home galaxy is the Milky Way which is a barred spiral galaxy which is over 100,000 light years across with a bar shaped central bulge of stars [Fig. 3.4].

Fig. 3.4 The Sun's position in the Milky Way.

Chapter 3. Twinkle, twinkle little star…

The Milky Way has four major spiral arms of stars, and several shorter ones. The Sun and the solar system are found approximately 28,000 light years from the galactic centre in its Orion arm which is between the major Perseus and Sagittarius arms. You can look it up on the internet. The Sun and its retinue of planets and other objects rotates around the galactic centre once every 240 million years.

Beyond the Sun, the star nearest to us is *Proxima Centauri*. Light needs to travel for almost 4 years and 3 months to reach us from Proxima Centauri, so this star is more than 4 light years from us. We learnt in Chapter 1 that a light year, the distance that light travels in a year, is about 10^{13} kilometres (10,000,000,000,000 km). Even the Sun is an exceedingly long way (150 million km or about 8 light minutes) from us. Beyond the Milky Way there are countless more galaxies at distances so vast that they are difficult to imagine. All are teeming with countless stars. Each star is a bubbling ball of gas, shining with its own light.

But what makes the stars shine?

How do stars shine?

Stars are primarily composed of hydrogen and helium; this fact was first proposed in 1925 by Cecilia Payne-Gaposchkin (1900-1979), a British-born American astronomer and astrophysicist. They are formed when a gas cloud composed of these gases plus some dust and perhaps other material from a nearby supernova explosion starts collapsing, either due to its own gravity or a shove given to it by the exploding supernova. As the cloud collapses, the pressure and temperature at its centre builds up, eventually becoming so high that gas atoms are stripped of their electrons, and their nuclei are pushed closer and closer together.

You may remember from Chapter 2 that the nucleus of a hydrogen atom consists simply of a proton which is a positively (+) charged particle. You may also recall that similar charges repel each other. So, when two nuclei of hydrogen come close together, they are repelled due to their like (positive) charges. However, as the pressure in the middle of the gas cloud continues to increase, it can eventually reach a point (when the temperature is more than 10 million K (10^7 K)) where the two positively charged hydrogen nuclei (protons) are forced together closer than a critical limit (10^{-15}m or 0.000000000000001m) where the *strong force* takes control and binds them tightly together, despite the *electromagnetic force* trying to push the similarly charged particles apart. The reaction also releases energy in the form of high-energy gamma rays. This process is called nuclear fusion. [Note: As we shall see in Appendix B, the Quantum Mechanics' tunnelling effect is also essential in bringing the protons together].

Nuclear fusion

[You may skip this section and return to it later]

Nuclear reactions due to protons combining changes every four hydrogen nuclei into one helium nucleus [Fig. 3.5] with its two protons and two neutrons. The process also produces particles called neutrinos (which have tiny mass and no charge) and positrons (which are antiparticles of electrons and so are positively charged). This process of converting hydrogen into helium is also called hydrogen burning.

[Note: The details of the interactions that convert hydrogen to helium, and other such processes are not covered in this book].

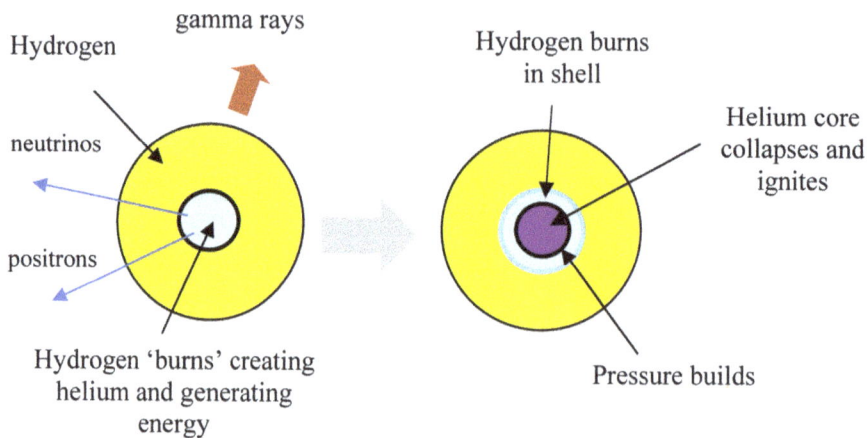

Fig. 3.5 Helium is produced by hydrogen burning in stars such as the Sun.

The helium produced in the core then collapses under pressure and ignites.

Where does the released energy come from? It turns out that if we weigh the nuclei of hydrogen and helium (yes, it is possible to weigh these tiny particles) we find that a helium nucleus weighs just a little less that did the four hydrogen nuclei with their four protons which created it. Remarkably, this tiny loss of mass precisely equals the mass of the neutrinos and positrons produced plus the energy of the gamma rays released, exactly as predicted by Einstein's famous $E = mc^2$ equation that we met earlier.

Stars therefore shine initially due to the energy released by converting hydrogen into helium. Hydrogen is their basic fuel. Each time hydrogen nuclei combine at the centre of the star, atomic particles (neutrinos and positrons) are released, as well as energy in the form of the very energetic gamma ray photons, and helium nuclei are produced.

48

Chapter 3. Twinkle, twinkle little star...

What happens to the neutrinos and positrons which are generated? The neutrinos react extremely rarely with other particles of matter, so they shoot straight through the body of the star and out into space. The positrons, being the electron antiparticles, disappear in a puff of pure energy photons when they collide with an electron. The gamma ray photons make their way to the surface, but ever so slowly. It has been calculated that in a star like our Sun, a photon released at the centre would be so battered about by the collisions with the electrons it meets along the way, that it would take two hundred thousand or more years for it to find its way to the surface and out into space. These collisions of the gamma ray photons with the electrons are very reminiscent of what happened in the Universe before the recombination, when photons were colliding with the electrons and could not escape.

The conversion of hydrogen into helium (hydrogen burning) is the basic reaction that sustains stars at the start of their lives. There is so much hydrogen in the Sun, for example, that the energy released will keep our star shining for almost 10 billion years. Some 4.5 billion years have passed since the Sun came into being, so it is roughly half-way through its life.

As the star is producing energy by the fusion of hydrogen nuclei, it is also producing helium. This adds to the amount of helium that was there in the cloud that produced the star. What happens to this helium?

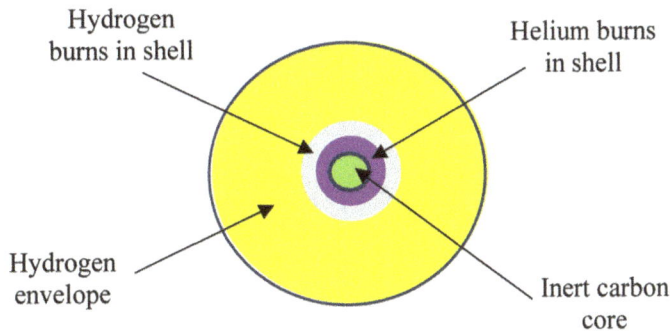

Fig. 3.6 Carbon is produced by helium burning in stars bigger than the Sun.

As the hydrogen 'burns' and is used up in the core, having been converted into helium, the core slowly contracts. This contraction releases energy which causes the hydrogen in the shell surrounding the core to ignite. The pressure from the hydrogen and the mass of the Sun above the core compresses the helium core which becomes even hotter. When the temperature reaches about 10^8 K (one hundred million degrees K) the helium starts to 'burn' due to reactions that push three helium nuclei together to form a carbon nucleus [Fig. 3.5, Fig. 3.6].

Later, if the star is big enough to generate the higher core temperatures needed, the helium and the new carbon nuclei can themselves react together to produce oxygen, and carbon burning. Similarly, for even larger stars helium and the new oxygen can react together to form neon (this is the gas that shines in neon-lights). The carbon, neon and oxygen fall into the core and burn one after the other, while helium burns in a shell above them [Fig. 3.7].

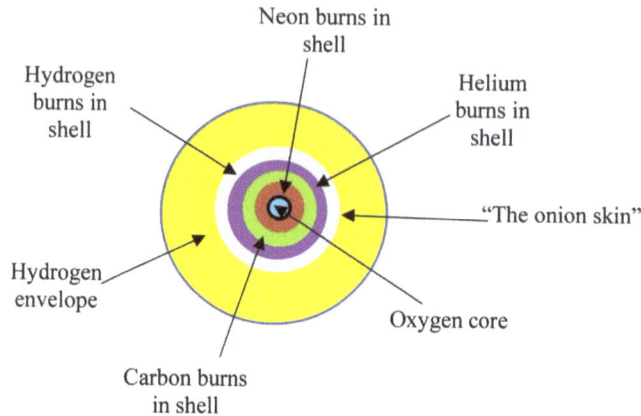

Fig. 3.7 Higher mass elements up to neon are produced in even bigger stars.

Do you see what is happening? Each stage of burning is producing the next element in a chain which then falls into the core, while the original reactions continue in a shell above. A good analogy that scientists often use to describe this is to think of the burning shells as onion-like layers with the newest formed elements being at the core.

During the fuel burning process, the pressure from the nuclear reactions heats up the contracting core, but also pushes outwards and balances the weight of volume of gas above the core of the star. This is what keeps the star stable and at the same size [Fig. 3.8].

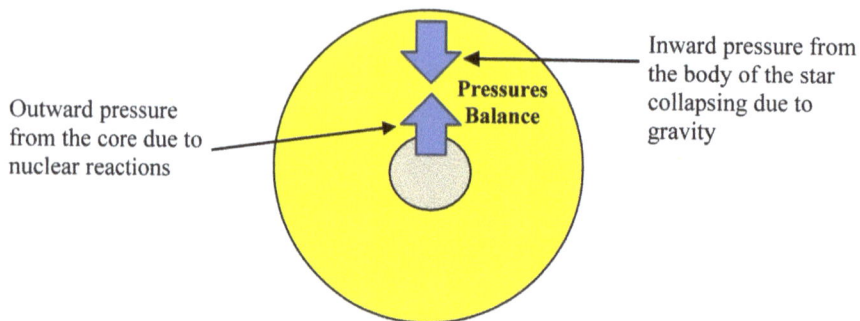

Fig. 3.8 A stable star.

Chapter 3. Twinkle, twinkle little star…

If there is enough gas in the star, that is, if the star is big enough, this process can continue till many more elements are formed. The onion layers are produced because of burning hydrogen, helium, carbon, neon, oxygen, and silicon one after another. The process continues till the core consists of iron. Then it stops. Why? We will answer this shortly when we see how stars die and come to realise how crucially important the answer is for us.

[End of optional section]

Let me divert a little and tell you about what I find most interesting in this whole scheme of things. It is this: we start with the normal matter in the Universe consisting almost solely of hydrogen and helium gas; the first stars form from clouds of these gases; the stars then produce *all* the other elements in their cores. If there had been no stars, the Universe would have consisted virtually only of hydrogen and helium and we would have had no oxygen, carbon, copper, iron, gold, or anything else that we can think of. There would have been no liquids, solids, or other gases - no rocks, no water, no air, and most importantly for us, there would have been no life.

Life as we know it is carbon-based. What this means is that carbon is the fundamental element that is needed for every form of life that we know of. This is because carbon has the capability of combining with hydrogen to form long-chain hydrocarbon molecules which are essential to produce the complex organic molecules necessary for life. Every single life-form that we are aware of – every animal, insect, reptile, tree, plant, and bacteria – is made of cells that are themselves made of carbon and hydrogen. Think about it. If there had been no stars, there would have been no carbon and no us (or carbon-based aliens for that matter). Every single cell in our body and that in every other living thing we know of has an element that was made in a star. You were made in a star. I was made in a star. We are truly star dust, in common with everything else in the Universe, other than hydrogen and some of the helium. I find this fascinating; I hope you do too. This was the realisation that made me want to learn about astrophysics and cosmology.

Now back to our story.

What difference does star size make?

The bigger the star the more fuel it has available to burn. The larger masses result in higher pressures and thus higher temperatures at the centre. These temperatures determine which elements can burn in the star centre to produce the next stage of burning. It turns out that stars which are between about one-tenth (0.1 times) to one-half (0.5 times) the mass of our Sun can burn hydrogen, but not helium. Stars between one-half (0.5) to eight (8) times the mass of the Sun will burn hydrogen and helium to produce carbon, oxygen, and neon, as we saw above. It needs stars between eight (8) and eleven (11) times the Sun's mass to burn carbon and the stars need to be bigger than eleven (11) times the mass of the Sun to burn oxygen, neon, and silicon and the other elements up to the production of an iron core.

It also turns out that the larger a star is, the more fiercely it burns its fuel and the sooner it dies. The smallest stars slowly fade over enormous lengths of time. Stars like our Sun last a few billion years, before they die shedding their gaseous envelopes and their cores slowly fading away. The largest stars have shorter lives and more spectacular deaths. They die in huge explosions called supernovae which can be visible across millions of light years. Fig. 3.13 shows the Crab nebula which is what remains following the explosion in 1054 of a star some 6,500 light-years from us. The supernova was bright enough to be observed during daytime from Earth.

Because there was so much gas around when recombination occurred, the first stars that were produced were huge, up to 50 to 100, possibly even 150 times the size of our Sun. These stars therefore, would have been extremely big and shone exceedingly brightly, for a remarkably short periods of time in comparison with their smaller brethren. Then they would have died spectacularly.

What about the little ones at the other end of the spectrum? How small can stars be? We have seen that stars form when huge gas clouds slowly collapse. We call an object a star if it can shine with its own light. That is, if it forms from a cloud that has enough gas to provide the pressure and temperature at its centre needed to start and maintain the reactions that burn hydrogen and light up the star. If you make a star smaller and smaller, a point comes when the hydrogen simply does not participate in any reactions because the pressure is not high enough to force the nuclei together. The minimum star mass at which hydrogen can truly ignite has been calculated at 0.08 times the mass of our Sun, in other words about one-twelfth of the mass of our Sun.

Interestingly, stellar (meaning star-like) objects only about 0.015 times the mass of our Sun (or some 66 times smaller than the Sun) can have reactions occurring in their cores that result in some radiation from their surface, though not at a level to qualify them as true stars. These are called *brown dwarfs* - dwarfs because they are small, and brown because they burn very dimly. In comparison with a brown dwarf, the largest planet in our Solar System, Jupiter, is about one-thousandth (0.001) the mass of our Sun, only about 15 times smaller than a brown dwarf. Even so, Jupiter does produce some internal heat.

The brown dwarfs are extremely hard to spot because they are so small and so dim. Observations using the Hubble telescope have shown that they are quite common., but we do not have a good idea as to how many they are. For a while scientist wondered whether the brown dwarfs were the missing mass of the dark matter. However, this is now discounted.

Chapter 3. Twinkle, twinkle little star…

How stars evolve

An especially important tool in establishing how stars evolve from their birth to the time they die is called the Hertzsprung-Russell diagram, or H-R diagram for short. As the name suggests it was developed through the work of two eminent scientists, Ejnar Hertzsprung (1873-1967) a Danish chemical engineer who became an astronomer, and Henry Norris Russell (1877-1957), an American astronomer. The H-R diagram [Fig. 3.9] is a graph showing how brightness of stars varies with their size and surface temperature. [Note that Blue is a hotter colour than Red, with White being the hottest. A white-hot poker is hotter than a red-hot one.]

The reason why the H-R diagram is so important is because the stars do not stay the same in terms of brightness and temperature from their birth to their death. And as we shall shortly see, the H-R diagram enables us to track the journey of the stars over their lifetime.

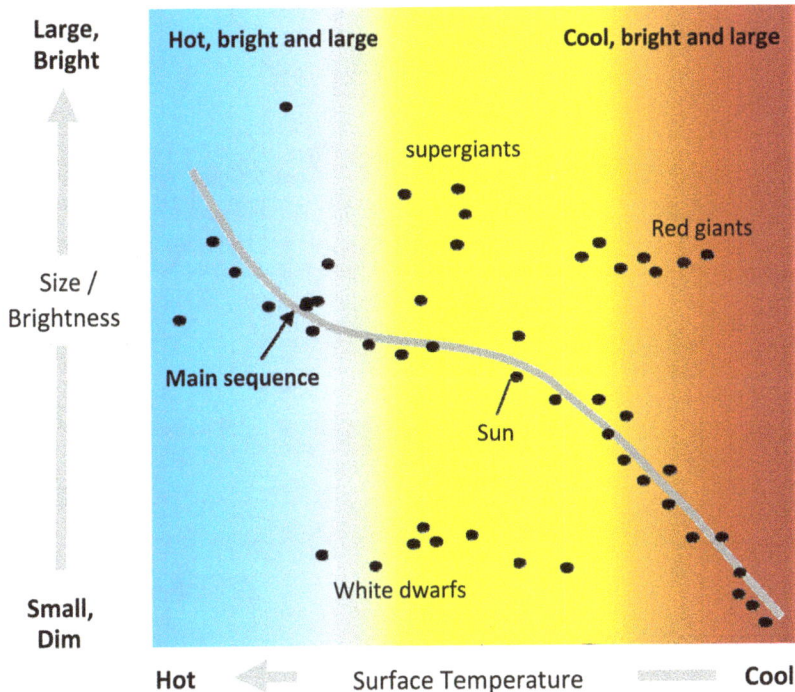

Fig. 3.9 The Hertzsprung-Russell (H-R) diagram.

Most of the stars are found in the band marked 'main sequence' in the diagram that runs from the top left part of the diagram down to the bottom right. In other words, the band ranges from very bright, hot, blue stars on the top left to the very dim, cool, red stars in the bottom right. The position of our Sun today is shown on the diagram. You will see that it is a very average star (with a surface temperature of about 5000 C) mid-way up the main sequence.

When the Sun was in its infancy cloaked in the gas cloud of its birth, it was bigger but less bright than it is now. Its position would therefore have been on the right of the main sequence.

Soon after their birth, stars develop outflows of gas we mentioned earlier that stream out at high speeds from it causing it to lose mass. The star then brightens and moves onto the main sequence; it has reached maturity. For a Sun-sized star, the move onto the main sequence would take about 10 million (10^7) years, while its life on the main sequence is expected to be around 10 billion (10^{10}) years. Thus, its childhood is only about one-thousandth of its mature life, which is short indeed. Larger stars reach their main sequence positions even faster. A star 15 times the mass of the Sun takes only about a hundred thousand (10^5) years to reach maturity.

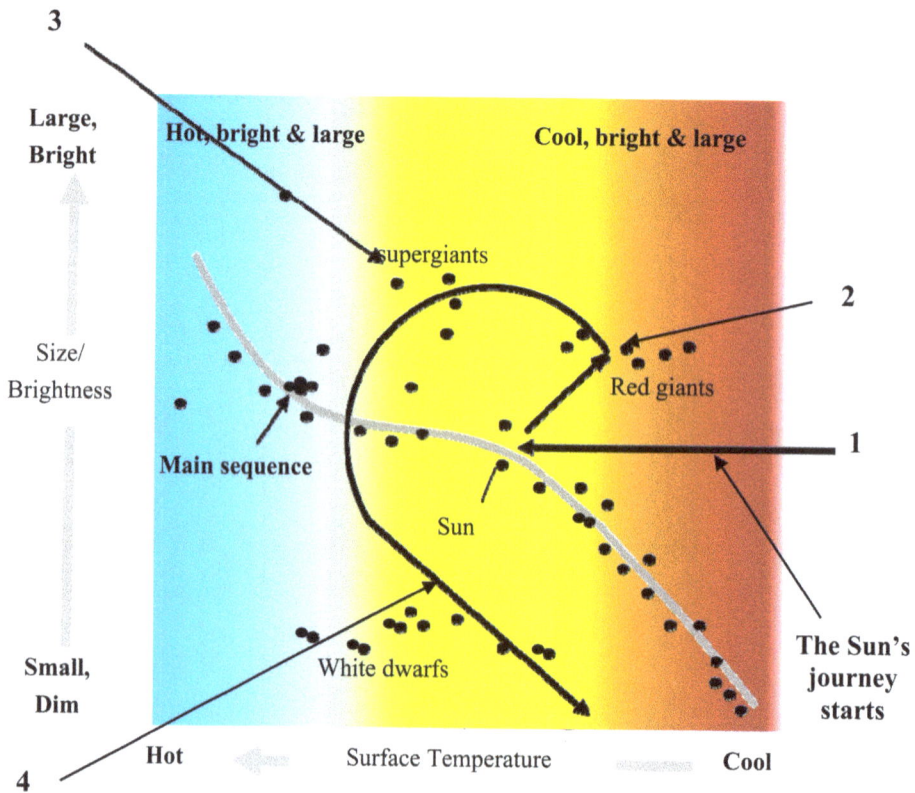

Fig. 3.10 H-R diagram showing the Sun's journey.
(see below for an explanation of the numbered arrows)

Chapter 3. Twinkle, twinkle little star...

Notes on Fig. 3.10

1. The Sun was born some 4.5 billion years ago. It moved onto the main sequence (the collection of long-lived stars along the pale-coloured line) as an average star.
2. In another 4.5 billion years it will start to die, and bloat up into a red giant, and move off the main sequence.
3. The Sun will shed most of its mass in a planetary nebula and move on its way to becoming a white dwarf.
4. The white dwarf will continue to fade, cooling slowly over many billions of years.

The next few pages tell the story of what happens when stars die. This depends on the size of stars, ranging from those of 0.08 solar masses which glow as brown dwarfs to the gargantuan stars of 50 to 100 solar masses which go on to form black holes. The story of how and why stars die will allow us to complete the Sun's journey on the H-R diagram.

How stars die

Stars die when they run out of fuel. But how they die and how long they live depend on their size. Big stars shine brighter and burn their fuel more quickly and so die at a younger age than the smaller ones. Let us see what happens by considering stars of different sizes.

The very smallest stars: up to 0.08 the mass of the Sun

The smallest stars are called brown dwarfs. They form out of collapsing gas clouds as do other stars. But brown dwarfs can barely be called stars. They are small and virtually invisible, glowing only very dimly. At the centre of the brown dwarfs the same hydrogen to helium reaction takes place as occurs in the Sun, but it proceeds very slowly. Therefore, the little fuel that a brown dwarf has lasts for a long time, with the result that these stars are the longest-lived stellar objects. Brown dwarfs will fade away slowly over billions and billions of years.

The small stars: up to 0.5 times the mass of the Sun

These have enough mass for hydrogen to undergo burning at normal rates, converting it into helium. But the subsequent phase of helium burning does not happen. The core temperature does not get high enough for the helium to ignite. The star will therefore shine while its hydrogen is burning but then fade. It will, however, live on for a long time as a dying ember.

The average stars: 0.5 to 5 times the mass of the Sun

The stars in this range, including the Sun (with a solar mass of 1), have a more interesting life cycle. They can ignite both hydrogen and helium. Once the hydrogen in the core is used up and replaced by the helium produced by the conversion of hydrogen, the next phase kicks-in and the helium core collapses since the hydrogen is no longer burning and providing the supporting pressure. The collapse compresses and heats up the core till the helium starts burning. Meanwhile the hydrogen continues to burn in a shell around the core. The burning helium produces carbon. However, the core temperatures are not high enough in these stars for carbon burning to occur.

We should note one important point. Each fuel that burns in a star produces an element with the next more massive atom, or more precisely, an atom with the next higher 'mass number'. Mass number is the number of protons and neutrons in the atomic nucleus. So, a helium atom with 2 protons and 2 neutrons in its nucleus has a mass number of 4, while a hydrogen atom with 1 proton in its nucleus has a mass number of 1. Therefore, hydrogen can burn to produce helium, but the reverse cannot happen. For simplicity, we will say that an element that has a higher mass number is more massive than one with a smaller mass number.

In the later part of the star's life, while the helium is burning, the star has two sources of energy, one from the helium burning in the core and the other from the hydrogen which continues to burn in a shell around the core. This produces an upward pressure on the gas in the star above the shell. The result is that the star expands some 10 times in diameter and becomes a supergiant.

The expanded star has the same mass but a much larger radius. The outer layers of gas are now further from the centre and therefore have a smaller gravitational force binding them to the star. The gas atoms drift away from the star and a solar wind develops blowing gas away from the star's surface. The star's energy gets spread over a much larger surface area of the bloated star and the star's mass reduces due to the solar wind. The surface temperature therefore drops, and with the drop the colour of the star changes. Depending on the mass of the star, the star may look blue, yellow or red. Our Sun which is yellow now will look redder when it is dying (the red end of the spectrum is at a lower temperature than the blue and the violet, you will recall). We call such a star a red giant.

There are many red giants that can be seen in the sky. One is Betelgeuse in the Orion constellation [Fig. 3.11]. Betelgeuse is so big that if placed at the Sun's location it will extend up to Jupiter's orbit. It is about 640 light years from us and is one of the very few stars which appears as a disk from telescopes on Earth rather than just as a point of light. The very bright star Rigel, also in the Orion constellation at the opposite corner to Betelgeuse, is a blue supergiant about 770 light years from us.

Another red giant is Aldebaran about 65 light years from us. Look for the three bright stars that form Orion's belt in the Orion constellation. Follow these up (or down in the southern hemisphere) to the fiery red Aldebaran which is about 40 times the diameter of our Sun. Aldebaran is in the constellation of Taurus the bull and sits at the eye of the bull.

Fig. 3.11 The Orion constellation.
Betelgeuse, the red supergiant, is at the top left.

Note: The Fig. 3.11 photo was taken in the Northern hemisphere. Orion appears 'upside down' if you view it from the Southern hemisphere (with Betelgeuse down on the bottom right, and the sword pointing up), as do all constellations common to the two hemispheres. This, of course, is simply due to Earth being a sphere - we have our feet on the ground and our head nearer the sky pointing in opposite directions in the two cases, e.g., in the UK and in New Zealand.

Becoming a red giant is the fate in store for our Sun. When it approaches the end of its life, it will expand till it almost reaches the Earth. It would have swallowed Mercury and Venus along the way. The Earth's oceans will boil away, its atmosphere will be stripped off and the land burnt to a cinder. Our home planet will not be a pleasant place to be on when it happens. However, this is still 4 or so billion years in the future. Hopefully, any surviving humans, in whatever evolutionary form they have attained, would have long since gone and settled on other planets, their new homes.

Stars which are two or more times heavier than our Sun can go through another phase as pressure builds up in their core.

In 1925 Wolfgang Pauli (1900-1958), an Austrian quantum physicist, put forward a rule (called the *Pauli exclusion principle*) which specified that there is a limit as to how closely electrons can be packed together simply through increased pressure.

When pressure is increased in the core, the electrons are forced closer and closer together and the temperature increases. However, according to the *Pauli exclusion principle*, as the pressure continues to increase, a critical value is reached called the *electron degeneracy pressure*. Pauli showed that if pressure is increased beyond this critical limit, the core pressure will remain the same even as the temperature continues to increase. This is because electrons in the core resist being crushed together beyond a certain point. These electrons are said to be *degenerate*. But such a situation cannot last, and something must give. In the star's case, there is an explosive release of pressure, the core expands and cools, stability returns, and the problem is removed. If the pressure increases again, the process repeats itself. The effect is that the red giants can become unstable and start to pulsate (bright, dim, bright, ...) with great regularity, as happens to Betelgeuse.

Note: at the time of writing, Betelgeuse had become so dim, that scientists were wondering whether it was coming to the end of its life. They eventually decided that is not so, and that the diming was due to a shadow on its surface cast by material that had been shed by the giant star.

Examples of such pulsating stars include the Cepheids which are an especially important class of stars. As we shall see later, they help us to measure distances to other stars and galaxies in space. In stars which are more than two solar masses, helium burning can start before electron degeneracy occurs. These stars do not pulsate but will still go through the red giant phase.

During the red giant phase, the stars may lose a lot of mass, equivalent of up to a fifth of the mass of our Sun, by shedding a shell of gas from their surface, which then expands and moves away from the star. This is called a *planetary nebula* [Fig. 3.12].

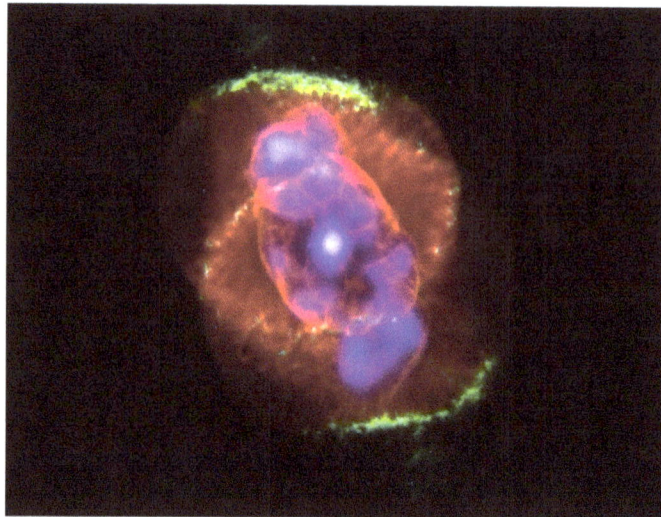

Fig. 3.12 The Cat's eye planetary nebula NGC 6543.
The nebulae are often incredibly beautiful objects in space.

Chapter 3. Twinkle, twinkle little star…

Planetary nebulae were so named because the shed shells of gas looked like hazy planets when viewed through the basic telescopes available at the time. The word nebula (plural nebulae) means a cloud. Today we understand better what the planetary nebulae are: the gaseous outer layers of their mass thrown out by dying stars. As we shall see later, larger stars that die in a Type II (pronounced 'Type two') supernovae explosion can also leave a nebula behind [Fig. 3.13].

When the outer envelope of the dying star has been lost into space, the core remains. It is called a **white dwarf** (because it is very bright and exceedingly small about the size of the Earth, but with the mass about that of our Sun). The internal gravitational pull on the mass of a white dwarf is supported by the pressure of the gas of degenerate electrons which we discussed above. Slowly the white dwarf will fade into a 'black dwarf'. But this will occur over many, many billions of years, so long, in fact, that the time since the Big Bang has not been long enough to produce any black dwarfs.

However, if the star is big enough, it may go beyond the white dwarf stage into even more exotic realms, as we shall now see.

The big stars: 5 to 11 times the mass of the Sun

Subrahmanyan Chandrasekhar (1910–1995), an American scientist of Indian origin, worked out that a white dwarf cannot be more than about 1.4 solar masses. This is called the *Chandrasekhar limit*. If the mass of the white dwarf gets above this limit, the electron degeneracy pressure can no longer support the mass of the white dwarf and the star collapses and then explodes catastrophically; such an explosion is called a Type Ia (pronounced 'Type one a') supernova.

Stars which are 5 to 8 solar masses have the capacity for their cores to produce oxygen from nuclear reactions between carbon and helium atoms, and neon from reactions between oxygen and helium atoms. They cannot go beyond this set of reactions though. Stars need to be bigger than 8 solar masses for their cores to be able to burn carbon to produce other elements such as magnesium, sodium, and neon.

As we well know by now, as each stage of 'burning' completes, its core contracts and its temperature increases. A new burning phase then starts in the core and in the shell above, leading to the pressure increasing and the star expanding further. When such a star finally dies after the various burning stages and reactions are complete, the result is once again a white dwarf with a mass less than 1.4 solar masses, supported by the electron degeneracy pressure.

There are other, more dramatic types of supernovae which occur when even larger stars die.

The giant stars: 11 to 50 times the mass of the Sun

We have seen that starting with hydrogen, stars 'burn' elements in their core to produce in turn progressively more massive elements – helium, carbon, neon, oxygen, …., till they run out of fuel. In general, the more massive the element that is produced by the nuclear reactions, the less is the energy that is released. In other words, the 'burning' reactions become progressively less efficient. A time comes when producing a new element costs more energy to produce it than the energy released by burning the fuel. This happens when we reach the element iron and occurs in stars which are more than 11 solar masses. Yet we know that we have elements on Earth that are more massive (strictly, have a higher mass number) than iron, such as silver, gold, and lead. So how are these elements produced?

Stars more than 11 solar masses can burn neon to produce oxygen and magnesium, burn oxygen to produce silicon, and finally burn silicon to produce elements up to and including iron. Because each 'burning' reaction gets progressively less efficient, each reaction must go faster than the last one to produce the energy necessary to balance the mass of the star and keep it from collapsing. As each burning reaction goes faster it also lasts a shorter time before the fuel is used up and the next burning must start.

Here are some fascinating numbers for you. They refer to a star of 25 solar masses.

Hydrogen burning lasts for seven million years, helium burning for half a million years, carbon burning for six hundred years, neon burning for one year, oxygen burning for six months and finally, silicon burning for just one day to produce the iron! *It takes just one day to produce all the iron that the star is going to produce.*

Burning iron is not efficient enough to produce any more massive elements. There the reactions stop. The radiation pressure from the core can no longer support the mass of the star. What happens then?

What happens is that the star blows up in what is called a Type II supernova. How and why this happens is explained in the section below; you are welcome to read it or to skip it till later.

How a Type II supernova occurs

[You may skip this section and return to it later]

When the silicon burning reactions stop, the iron core is no longer producing the energy to balance the mass of the star. The core contracts due to gravity leading to the electrons at the centre of the core becoming degenerate. This results in the electron degeneracy pressure which tries to but fails to halt the collapse of the core. In stars of more than 11 solar masses, the core is more than the critical 1.4 solar masses of the *Chandrasekhar limit*.

The collapse continues and the iron nuclei gets squeezed closer and closer together till the temperature reaches about 10^{10} K (ten billion degree Kelvin). At this temperature, the core nuclei start to disintegrate into protons, neutrons, and helium nuclei. These absorb some of the energy in the core and the collapse really speeds up and goes supersonic.

The electrons are now forcibly absorbed into the protons to produce neutrons. This further reduces the pressure and speeds up the collapse even more. *The core now reaches the similar density to that of an atom and is falling at up to 23% the speed of light*. The core temperature is now 10^{11} K (one hundred billion degrees Kelvin!). The neutron degeneracy now kicks in and a new pressure comes into play due to this. The collapse of the core suddenly comes to a halt and reverses.

However, the rest of the star is falling due to gravity at speeds of up to 70,000 km (45,000 miles) *per second*. This must stop suddenly. A huge shock wave is generated going upwards. The star blows up. This is a *Type II supernova*. What remains is a cloud of gas and dust in space, often shining in brilliant colours in both the visible and non-visible parts of the spectrum [Fig. 3.13].

Fig. 3.13 The Crab nebula – remnant of a Type II supernova.
The result of the explosion in 1054 of a star 6,500 light years from Earth.

[End of optional section]

The collapse of the core and the star blowing up in a Type II supernova all happen in a very few but extremely important seconds. *It is during these few seconds of the explosion that elements beyond iron, such as gold and silver, are created by the extreme conditions of temperature and pressure encountered by the material being expelled.*

The explosion produces an unbelievable number of high energy particles which stream away from the dying star. The sudden release of energy results in the star becoming more than a hundred million (10^8) or so times brighter than normal. *For a while, the supernova may become brighter than the rest of the billions of stars in the galaxy in which it lived. It becomes so bright that it can be seen halfway round the visible Universe*! The explosion producing the Crab nebula was recorded by Chinese astronomers on Earth even though it was 6,500 light years away.

Then the star dies leaving behind a nebula and a small, very dense object known as a *neutron star*. It is no bigger than about 20 km, or 12 miles or so in diameter, yet has a mass of between 1.4 and 5 solar masses. Imagine three or more Suns squeezed into a ball the size of a town. If it were possible to weigh this amazing object, each teaspoon worth weighs many billion tons. We will learn more about neutron stars in the next chapter.

But if the star is larger still, even a neutron star is not the limit in our story.

The gargantuan stars: 50 to 100 times the mass of the Sun

One of the pressures found in a star that we have not mentioned yet is the *radiation pressure due to photons* (light particles) that are released from the star. This pressure is extremely small in normal stars but can become significant as the stars get bigger and the temperature gets higher. When a star is about 100 solar masses (100 times as massive as the Sun), the radiation pressure is high enough to destroy the star. This sets the upper limit to how big a star can get. In any case, the life of a 100 solar mass star is extremely short due to the rapid depletion of its fuel.

The result of the death of the most massive stars is like that of the 11 to 50 solar mass stars: a supernova. However, in this case, the final product may not be a neutron star. The collapse of the core may not be stopped even by the neutron degeneracy pressure. If the core is more than about 5 solar masses, the neutron degeneracy pressure will be unable to stop the collapse.

The core continues to collapse, but into what? We do not truly know what is at the centre of such a star at end point of its collapse. What we are certain about is that the final remnant of the collapse will be a **black hole**. This is a strange and wondrous object with unbelievable properties.

We will look at black holes in more depth in the next chapter.

Chapter 3. Twinkle, twinkle little star…

How do we know?

Measuring distances to the stars and galaxies

We have talked of our nearest star Proxima Centauri as being over 4 light years from us. We have said that the Andromeda galaxy, the nearest major galaxy to us, is 2.5 million light years away. Others are many, many times more distant. But how do we know? How do we measure such vast distances? There are several different techniques we adopt. Each one is like a rung of a ladder, it serves its purpose up to a certain distance then another technique, another rung, needs to take over. Let me tell you about some these techniques.

The first way is by using the principle of *parallax*. What is parallax? Well try a simple experiment. Put your forefinger pointing up about 30 cm in front of your eyes. Then close one of your eyes, say the left one, and line the finger with a distant vertical object such as a telephone post outside or the edge of wall if indoors. Now close your right eye and look at your finger with your left eye. The finger would appear to have moved to the right. This is parallax and occurs because your eyes are a little way apart. Next repeat the experiment, but this time with your arm outstretched. You will find that your finger again appears to move, but not as much as before. By knowing how far apart your eyes are and by measuring how far your finger appears to shift it is possible to calculate relatively simply the distance from your eyes to your finger.

The same technique can be adapted to measure the distance to the stars. However, as we noticed when we moved the finger further away, the greater the distance of the finger from your eyes, the less was the shift to the right.

Now we know that the stars are far away, so trying to use parallax with your eyes will not work. Try it out on a starry night; the stars will not move when you swop your eyes to look at them. The only way to do it is to look at the stars through eyes that are much, much further apart. But how do we do that?

The trick is to look at a star and note its position precisely against the other stars in the sky [Fig. 3.14]. Then wait six months for the Earth to move halfway in its orbit around the Sun. That night we look at the same star and note its position against the other stars. We will see that it has shifted against stars that are much further away than the one we are measuring. Since we know the distance between the two points in the Sun's orbit from where we took the measurements (we know the Earth's orbit very accurately), and because we also know how much the star appears to have moved, we can work out its distance from Earth.

I am sure you have already realised that this will only work for relatively nearby stars. If the stars are extremely far away, their shift is again not noticeable even across the Earth's orbit. We can however improve the accuracy and precision of our measurements dramatically with telescopes in satellites. The Hipparcos satellite (High Precision Parallax Collecting Satellite) sent up by the European Space Agency (ESA) in 1989 measured the parallax of 118,000 stars at distances of up to 30 light years. The 2014 ESA satellite Gaia extended the distance to thousands of light years.

But what about stars and galaxies even further away?

We know that the further away an object is from an observer, the less bright it appears to the person watching. By comparing actual brightness and the observed brightness of an object we can work out the distance to the object. We saw earlier that a Type Ia supernova always explodes when its mass exceeds the Chandrasekhar limit of 1.4 solar masses, and therefore every Type Ia supernova explodes with the same brightness, which we can calculate knowing its mass. When we spot a supernova occurring in a far-away galaxy, we can measure its observed brightness from Earth. Then using the supernova's actual and observed brightness measurements, we can calculate how far the supernova is from us, and therefore how far is its parent galaxy.

Phenomena such as Type Ia supernovae which shine with a known brightness are called *standard candles*. We have other standard candles in space. Among these are Cepheids, the pulsating stars we mentioned earlier in the chapter when discussing the electron degeneracy pressure and the red giants.

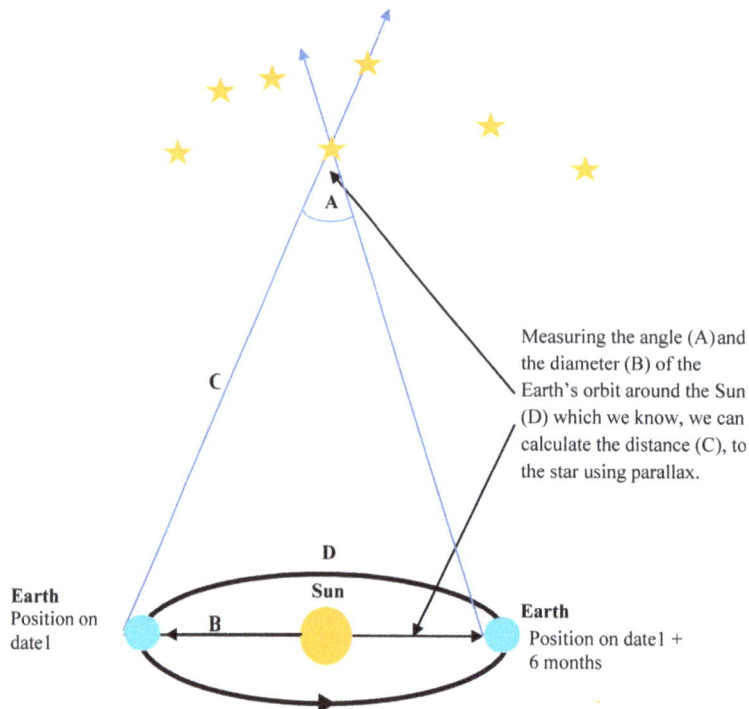

Measuring the angle (A) and the diameter (B) of the Earth's orbit around the Sun (D) which we know, we can calculate the distance (C), to the star using parallax.

Fig. 3.14 Parallax, and its use in measuring distances to nearby stars.

Chapter 3. Twinkle, twinkle little star…

Cepheids are among the group of so-called variable stars whose brightness varies with time. Cepheids pulsate, grow bigger and then smaller as their radius changes with time. This leads to their brightness varying with time with great regularity. Like for all the other waves, the time between two peaks of a Cepheids' brightness curve is called its period [Fig. 3.17(a)]. In other words, a Cepheid star with a period of 5 days will brighten to a peak then dim and brighten again every 5 days. For a Cepheid, this period can range from about 1 to 100 days.

There are two other things significant about Cepheids.

One is that the longer the period of a Cepheid is, so proportionately brighter is the star [Fig. 3.15 (b)]. For example, a Cepheid with a 10-day period will be twice as bright as one with a period of 5 days. Thus, if we know the brightness of the 5-day Cepheid, we will know the brightness of every other Cepheid.

The distance of nearby Cepheids can be measured using some of the other techniques and used to calibrate the brightness-period relationship.

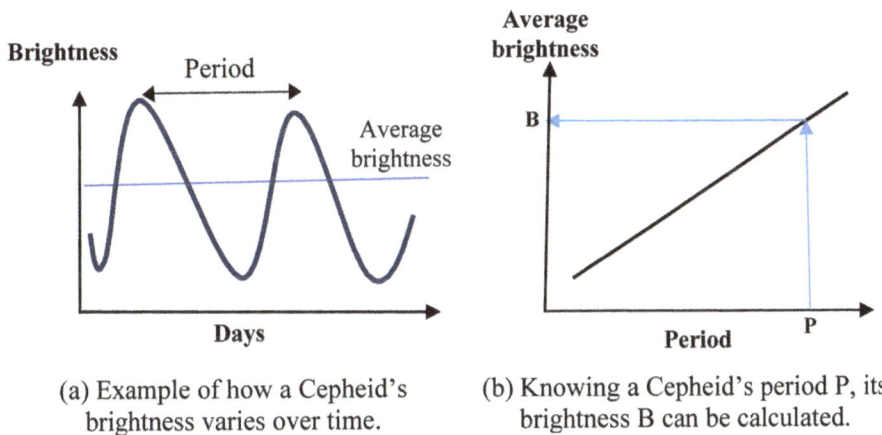

(a) Example of how a Cepheid's brightness varies over time.

(b) Knowing a Cepheid's period P, its brightness B can be calculated.

Fig. 3.15 Calculating a Cepheid's brightness.

This remarkable characteristic about Cepheids was discovered by Henrietta Leavitt and is called "Leavitt's Law". Leavitt was an American astronomer who worked at the Harvard College Observatory for many years as a "computer", analysing photographic plates to measure and catalogue the brightness of stars. She discovered the law relating to variable stars (over 2400 such stars are credited to her). Without her pioneering work, it would have taken much longer for distances to far galaxies to be accurately measured. Hubble used Levitt's work after her death to enable the Hubble relationship between the redshift and distance to be precisely realized. This relationship, as we have seen in Chapter 1, was the key to demonstrating that the Universe was expanding. Sadly, Henrietta Leavitt died before she could be nominated for a Nobel prize as was planned. She is one of few women astronomers to be recognised for their work.

Another interesting fact about Cepheid stars is that they are found all over the Universe. When a Cepheid is found, its period can be measured by examining its light curve. This will give us its actual brightness using the period-brightness relationship found by Leavitt. We can then measure its brightness as seen here on Earth. These two measures of brightness will tell us how much the light from the Cepheid has dimmed due to its distance of travel, and therefore tell us how far the Cepheid is from Earth. By observation we can identify the galaxy in which the Cepheid occurs, which tells us the distance of this galaxy from Earth.

Chapter 4

The beasts in the sky...

black holes and other nasties

The Universe has many examples of extreme energy. Some we have met already. We have seen that as the larger stars die, they do so in enormously powerful explosions called supernovae. These range from the Type Ia supernovae, which occur when the mass of a white dwarf exceeds the Chandrasekhar limit, to the much larger Type II supernovae that result in neutron stars or even black holes being formed following the death of the original star.

This chapter is about what follows the catastrophic death of exceptionally large stars. We will revisit supernovae, briefly. We will look in more depth at some of the awesome consequences of supernovae explosions, including black holes, pulsars, and gamma ray bursts. We will also look at other spectacular events that occur in the Universe: when galaxies collide and when white dwarfs, neutron stars and black holes crash into each other. We will learn of the phenomenal energy that these events release which enables us to see them right across the Universe.

Supernovae: a review

We have seen that the larger stars die in a spectacular fashion, in explosions called a Type II supernovae. These supernovae are extremely powerful and are valid candidates to be included in this chapter on their own merit. What results from the supernovae are highly extreme objects: neutron stars and black holes. We shall see the mayhem that these bodies can cause. But first let us remind ourselves of how supernovae are created.

A star shines due to the radiation created by the nuclear reactions in its core. The same radiation also produces the radiation pressure that supports the star against the inward pull of gravity on the mass of the star. The star collapses when the nuclear reactions can no longer generate enough radiation pressure to support its mass. This can happen when the fuel that is used in the nuclear reactions runs out. It can also happen when the new elements which are created by the reactions in the core cannot burn efficiently enough to generate the radiation necessary to support the mass of the star. You may remember from our earlier chapters that this can occur in the most massive stars after iron is created in the core of the star.

The collapse of the star when it occurs is very sudden, taking only a matter of minutes. The core is crushed as the whole mass of the star falls in on it, and it explodes in a supernova. For a star between 5 and 11 solar masses, the collapse is stopped by the neutron degeneracy pressure. You will remember that this is the pressure which results from the neutrons in the collapsing core resisting being pushed any closer together. This neutron pressure balances the mass of the in-falling gas and stops the collapse. The result is a neutron star.

For larger stars, even the neutron degeneracy pressure cannot stop the collapse. The collapse continues and we end up with the most extreme object possible: a black hole.

The smaller stars, up to 5 solar masses die more sedately, shedding their outer layers of gas to become planetary nebulae. These stars include those that are the size of our Sun which, of course, has a solar mass of 1. The collapse of the core leads to the electrons being squeezed till they reach a limit beyond which they become degenerate and resist being pushed together any closer.

This electron degeneracy pressure stops the collapse of the core, and what is left behind is a white dwarf. The white dwarf does not have any nuclear reactions to make it shine and it fades away over the eons. However, it carries a deadly capability. If it somehow manages to get more material added on to its mass till the mass exceeds the 1.4 solar mass Chandrasekhar limit, the electron degeneracy pressure can no longer support the white dwarf's mass and it explodes as a Type Ia supernova.

So interestingly, while white dwarfs do not form from a supernova, they can *cause* a supernova. As we have already noted, these Type Ia supernovae are extremely useful to science: they all explode with the same mass and therefore the same power and brightness which enables them to be used as standard candles to measure distances to far away galaxies.

So much for the supernovae explosions powerful as they are; we are now interested in what follows the supernovae. Let us therefore go on to find out more about what supernovae explosions can leave behind - the black holes, and other extreme objects of the Universe.

Black holes

A black hole is an object so dense that nothing can escape from its gravitational pull; nothing at all. Not even light. It is the blackest thing imaginable. It is a hole in space, a bottomless pit in our Universe, one that swallows everything unfortunate enough to come into its range but spits nothing out. It is a one-way exit from our Universe.

We know that all objects have a gravitational pull depending on their mass. Anything that is on the surface of an object in space needs to have a speed greater than the 'escape speed' for the object to escape the pull of the object's gravity. For example, if we try and go to the Moon from the Earth, we need to travel faster than some 11 km (7 miles) per second, or about 25,000 miles per hour, to escape the clutches of the Earth's gravity. On the other hand, it takes only a speed of 2.4 km (1.5 miles) per second, or about 5,400 miles per hour, to escape from the smaller Moon, with its smaller gravity, to return to Earth. On a black hole the escape speed is the speed of light which nothing can exceed, not even light itself! So, nothing can ever escape the pull of gravity of a black hole. Any object falling into a black hole is destined to be crushed out of existence.

I said, 'on a black hole'. But, of course, a black hole has no surface. The distance from the black hole's centre to where the escape speed reaches the speed of light is called the *Schwarzschild radius*, named after Karl Schwarzschild (1873-1916), the German physicist and astronomer. The area defined by the Schwarzschild radius is known as the black hole's *event horizon*. We can consider the event horizon as the black hole's 'surface' [Fig. 4.1].

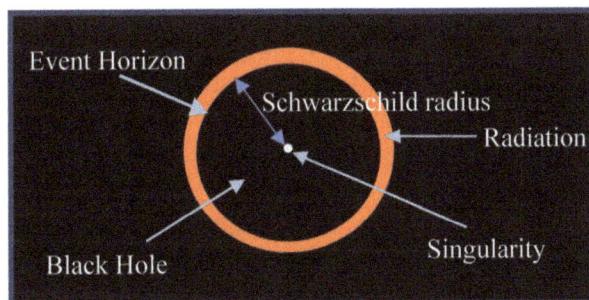

Fig. 4.1 The structure of a black hole.

There is an excellent analogy for a black hole: a waterfall. If you have seen a waterfall from above, particularly a large one, you would have seen that the river flows quite gently as it approaches the falls, and then speeds up as it nears the drop itself. If there is a swimmer in the river, he or she will float along unsuspecting, easily swimming against the current, till a point-of-no-return is reached. From that point on, there is no escaping the current and the deadly drop for the swimmer. It is similar in the case of a black hole. You can imagine the point-of-no-return to be the event horizon, the boundary where the black hole (or the waterfall) begins.

Let us repeat the startling facts: nothing can travel faster than light; thus, anything that is in/on a black hole can never hope to reach the black hole's escape speed, which is the speed of light, so nothing that falls within the black hole's event horizon can escape from its clutches, ever. Consequently, if when a star collapses its radius becomes less than its Schwarzschild radius, it is doomed to become a black hole. It will disappear forever. Anything that comes within the critical event horizon distance of the new black hole is also gobbled up. Since the Schwarzschild radius of a black hole depends on its mass, the more mass the black hole swallows, the bigger the black hole gets, the bigger its event horizon becomes, and the larger its gravitational attraction gets.

We do not know what happens to the material that is swallowed by a black hole. We have no idea what is at the centre of a black hole. Here the falling matter is crushed to nothingness, its mass becomes infinite and its size becomes zero. We call this centre a *singularity*. We do not know what it is. Our physics breaks down at these extreme limits. None of our theories work: neither Einstein's relativity which deals with the extremely large (gravity), nor quantum mechanics which deals with the exceedingly small (space). It is a bugbear for science. We need something new. There are a lot of scientists busy trying to resolve the problem. Until they do, we will not know what is going on within this beast in the sky.

To confound matters further, some black holes can have an electric charge, and some can be spinning. These factors give rise to even more complicated scenarios, including an inner horizon, a ring-shaped singularity at the centre, regions inside the black hole where space and time flip: in our normal world we can only travel forward in time, when the space and time flip, we can only travel forward in space, towards the centre! You will be pleased that we do not plan to probe any deeper into these concepts in this book.

What we do know is that black holes exist. They can be found in different locations and be of different sizes. Let us do a brief survey of the beast.

Chapter 4. The beasts in the sky…

Black holes form when massive stars die

We have seen that black holes may form as the result of the death of stars of over 11 solar masses. These black holes can form during the star's dying throes when the core collapses. If the star is large enough, the collapse is such that it cannot be stopped by either the electron- or the neutron-degeneracy pressures. The core disappears completely and what results is a black hole. As an example of the size of black holes formed by collapse of stars, the Sun would need to collapse to less than 3 km (2 miles) in radius for it to become a black hole. But since it has a mass of 'only' one solar mass, it is not big enough to ever collapse to this size and so will never form a black hole (phew). A more typical size of a black hole would be a radius of about 30 km (19 miles).

Black holes at the centre of galaxies

Most galaxies are believed to have a black hole at their centre. These black holes tend to be supermassive: many millions, even billions of solar masses. How and when these formed is still open to debate. Perhaps they came about when the stars were first forming and collecting to form galaxies. The huge early stars could have collapsed and produced colossal black holes. These huge black holes could have themselves collided to produce even more massive black holes. The black holes could have grown further by consuming the large amounts of gas which was around in the early Universe.

Mini black holes

Mini black holes are also theoretically possible. Some scientists believe that they may have formed at the time of the Big Bang. Stephen Hawking (1942-2018) the renowned British mathematician and cosmologist did a lot of work in this field. He was famous for having discovered the *Hawking radiation*, which is the radiation that a black hole emits under certain conditions. Since radiation is energy, which is equivalent to mass, the Hawking radiation means that a very tiny black hole can, in theory, evaporate away. We talk more about the Hawking radiation in Chapter 7.

Some bizarre effects of a black hole

Suppose two astronauts in the far future leave a spaceship (suitably dressed for the space walk, of course) which is orbiting a black hole at the centre of our galaxy, to investigate the supermassive beast more closely. They do not feel any different to normal while approaching the event horizon. They see a glow around the black hole itself as the stars near the horizon appear to be stretched along the curvature of the hole due to the black hole's gravity bending their light around the hole. [Note: this implies that if the astronauts look down the light, they will see the back of their head!] If the astronauts look back at their ship, it will seem to be orbiting the black hole faster and faster. On the other hand, to their colleagues left behind on the ship, the astronauts appear to be acting in slow motion, all their actions have slowed down. The two space-walkers' watches still work properly and agree with each other's. But if they could compare their watches with those of their colleagues in the spaceship, they would find that their own are running slower.

The astronauts do not realise it, but unwittingly they are drifting closer and closer to the event horizon itself. Without noticing it they cross it. They feel no different. But to their colleagues on the ship, they seem to have stopped, frozen in time, and are not moving. However, to the astronauts their ship has become a blur as it speeds around the black hole. After what they consider an unremarkable investigation during which they have discovered nothing strange, they fire their thrusters to return to the ship. The thrusters fire, but the astronauts do not move. They cannot return, and their ship has disappeared. Slowly they fall deeper into the hole, there is nowhere else they can go. If they are 'standing' feet towards the centre of the hole, they are stretched spaghetti-like, as the immense gravity of the black hole tugs more at their feet than their head. The end is nigh. There is no escape. To their colleagues still on the ship, they remain frozen at the event horizon. Weird or what?

Einstein in his *theory of general relativity* showed that gravity affects time. The larger the gravitational force, the slower the time moves. This is not just that watches go slower, but time itself slows down. What this means is that if you were in a place of high gravity your time would be going slower than the time for your friend who is in a location of lower gravity. You will be growing older slower than your friend. This is the anti-aging elixir man has been looking for! Unfortunately, the effect is tiny; but it is measurable. It is called *time dilation* and was predicted by Einstein in his *theory of special relativity*.

Many experiments have been done to test time dilation; they all confirmed it truly exists. If you take an extremely accurate clock to the top of a mountain and compare it with a clock at sea level you will find that the clock at the peak is running faster – the gravity is lower on a mountain top, which is further from the Earth's centre, than at sea level. The clock slows down as it is brought down to sea level. Understanding this effect of gravity on time is crucial when developing applications such as GPS (Global Positioning System) which enables our smartphones to know where we are. The software on satellites which are used for GPS have built-in corrections based on Einstein's theory to take into account the time dilation between the Earth and the satellite.

Gravity also affects space itself. Distances, and shapes such as the angles in a triangle, are no longer the same in a place with extremely high gravity as they are in normal space. Gravity bends space. In fact, gravity affects what Einstein called *spacetime*. Einstein showed that space and time are irrevocably linked together. We shall look more into this in Appendix A. For now, it is enough to know that what is happening near the black hole event horizon is that space is becoming more and more curved due to the huge gravity of the black hole and time is slowing down, as it did for our astronauts as they approached the black hole's event horizon. At the event horizon itself, time stops and the space curves completely. The door shuts. To those far from the event horizon in normal time, the astronauts are frozen in time. For those inside the event horizon all is lost.

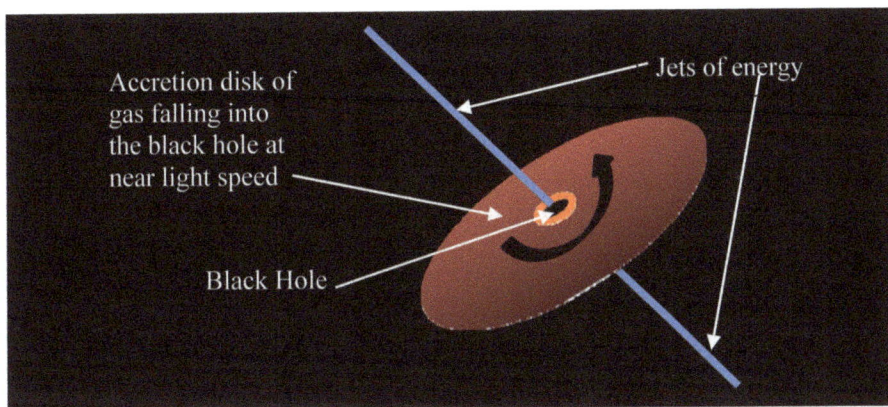

Fig. 4.2 Gas swirls around a Black Hole before falling in.
Jets of energy shoot from the poles.

We chose the large black hole at the centre of our galaxy for the astronauts' story. If we had chosen a smaller stellar-sized black hole formed by the collapse of a star, the end for the astronauts would have been quicker and nastier. Due to the smaller radius of this black hole, the difference in gravity between their feet and head would have been many times larger, and their end would have come quicker and been more dramatic.

How do we know black holes exist?

Black holes cannot be seen. For many years many people, including Einstein himself, thought they were impossible to occur in real life, even though his theory predicted them. He thought them far too bizarre to be possible.

We now know that they really exist. But how do we know this?

By theoretical analysis: Einstein's work on gravitation and other studies of stars, their formation and death, tells us that black holes could exist.

By observing jets from some galaxies: even though we cannot see a black hole itself, we can observe the effect it has on the surroundings. When gas falls into a black hole - either because a gas cloud has come too near it, or by gas being sucked from a star that is orbiting the black hole, [Fig. 4.2] it does so by forming a disk (called an *accretion disk*) of material that spins at tremendous speeds around the black hole before plunging into the event horizon, rather like water does as it pours down a plug hole. The enormous gravitational pull of the black hole speeds up the spinning particles to near the speed of light. The extreme friction between the particles heats them up as they orbit the black hole so that they glow and radiate enormous amounts of energy in the form of jets shooting from the poles of the black hole [Fig. 4.2].

The jets can be extremely powerful and extend for distances of many light years out of the galaxy. If they shine in our direction and are powerful enough, we can observe them through telescopes on Earth. We know pulsars (see later in this chapter) also have beams of energy shooting out from their poles. But we know of no means other than a black hole that can create beams of such power as are found to be coming from some galactic centres at such distances. Such galaxies are called quasars. Below, we talk a little bit more about quasars.

There are no jets from the black hole at the centre of our Milky Way galaxy. But 'our' black hole is probably just sleeping. If a star comes too close, or if a gas cloud drifts into its zone, it may yet awaken and shoot jets into space. That will be a sight to behold. Luckily for the Milky Way the jets will shoot out at right angles to the galaxy's disk, so most of the stars should not be in any danger, including the Sun, if it still exists at that time that is.

By observation of our own galactic centre: when an object orbits around another, the nearer that that the orbiting object is to the centre the faster it will go around. For example, in the case of a planet going around a star, the nearer the planet is to the star, the faster it will orbit it. Thus, Mercury goes around our Sun faster than any other planet, and the Earth goes around the Sun faster than does Mars which is further away from the Sun.

When we look at the centre of our own galaxy, the Milky Way, we find several stars which are circling around a point in space in very tight orbits. The stars are going around fast enough for them to be filmed (you can see this on the internet) and for their time-to-orbit to be measured. The time that an object takes to orbit another object, and the distance between the two objects, can be used to determine the mass of the central object. When we do the exercise for the Milky Way, we find that at the centre of the orbits of all the stars at the hub of our galaxy is an object we cannot see, but which has a mass equal to 4.3 million solar masses. We believe that this is a black hole; nothing else fits the bill of being so massive, yet invisible. While the size of our black hole appears huge, other galaxies are thought to have black holes at their centre which are billions of solar masses, such as the black hole at the centre of M87 galaxy [Fig. 4.3].

Chapter 4. The beasts in the sky…

By taking a picture: On Wednesday, 10 April 2019, a project known as the Event Horizon Telescope (EHT) released an image of a black hole [Fig. 4.3, below]. This was a remarkable achievement of a hugely intricate and complex project. It involved setting up telescopes in 8 locations across the face of the Earth, ranging from Arizona, Spain, Hawaii, Chile, Mexico, and the South Pole.

All these telescopes were trained on the same point in space looking at a huge black hole which was believed to exist at the centre of the M87 galaxy. The black hole is 38 billion kilometres (24 billion miles) across, or at least its shadow that we see is, and is some 6.5 billion times as massive as our Sun! The event horizon itself is about 5.5 times smaller than the diameter of the shadow, or about 7 billion km (4.4. billion miles) across. It is 55 million light years from us. The scientists were examining two black holes, the huge one in the M87 galaxy, plus the one at the centre of our own Milky Way galaxy called Sagittarius A*, which is *only* 4.3 million solar masses and some 26,000 light years from the Earth.

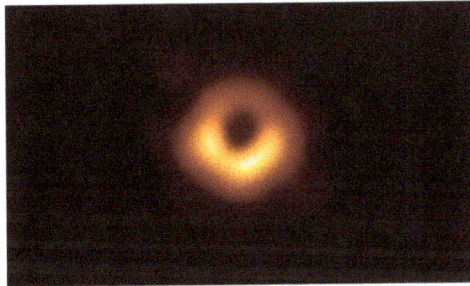

Fig. 4.3 The Black Hole at the centre of M87
galaxy 55 million light years away.

The M87 is the first black hole to be imaged. The eight telescopes were linked together such that they acted as a single instrument virtually the size of the Earth. They were synchronised to capture the image of the black hole at the same instant. This was critically important if a consistent image was to be captured. Finally, when all was ready, the scientists just had to wait for clear skies at all the 8 locations at the same time. The result though was worth all the effort.

We have known that black holes exist because science in the form of Einstein's theory told us so. But the final proof was always going to be to see the 'unseeable'. Einstein was proved right once again.

You will see from the picture that one half of the ring of dust around the black hole is brighter than the other half. The reason is that because the ring is rotating at near the speed of light, the ring particles moving toward the Earth appear brighter than the particles moving away from us. No doubt that in the future we will see images of other black holes including the one at the centre of our own Milky Way.

Neutron stars and pulsars

We saw in the last chapter that when giant stars of some 11 to 50 solar masses die, they do so in a tremendous supernova explosion. The explosion of this intensity is the result of the core collapse of these huge stars. The core collapses so violently that even the electron degeneracy pressure that supports a white dwarf produced by the death of a smaller star is unable to hold it back; the collapse may stop only when the neutron degeneracy pressure comes into play.

Neutron degeneracy develops when the (negatively charged) electrons in the atoms are forced into the (positively charged) protons at the nucleus of the atom to produce (neutral, no charge) neutrons. All the neutrons are forced together till they cannot be squeezed in any more due to the Pauli *exclusion principle* which we talked about earlier. The result is a **neutron star**.

The supernova that produces a neutron star is remarkable enough resulting as it does in the release of energy that makes it shine more than 100 million (10^8) times brighter than the star was shining before it exploded. But neutron stars have many other very extraordinary characteristics.

The resulting object is amazingly small, being only about 20 km (or 12 miles) in diameter, about the size of a town, yet is massive at between 1.4 and 5 solar masses. It is extremely dense, so dense that a teaspoon of the neutron-star material would weigh many billion tons. Its gravity is many hundred billion times more than that of the Earth's. Its temperature is many million-degree K. It has a magnetic field that is tens of millions of times larger than that of the Earth's. It spins at rates of up to hundreds of times per second. And it shoots out narrow beams of high energy radiation from its poles, like the jets from a quasar we met earlier [Fig. 4.2], though it is far less powerful. Everything on the star would be squashed to its atomic particles, burned to a sizzle, and blown to smithereens. Not a place one would want to visit.

The energy beams from a spinning neutron star are so powerful that they are detectable over millions of light years. If they are pointing towards the Earth, they are powerful enough to be seen by telescopes here. The spin of the neutron star is very rapid and very precise, so much so that when they were first discovered in 1967 – by Jocelyn Bell Burnell (1943-) when conducting research in Cambridge for her PhD – it was thought that the rapid pulses could be messages from intelligent alien life (the star was initially called *LGM-1*, short for *little green men-1*). Once the true nature of the pulses was realised, it was called a **pulsar**.

Pulsars can spin at different rates. One has been found that spins at more than 650 times per second. This spinning rate means that its surface is going around at an amazing speed of over 40 thousand kilometres per second (over 25 thousand miles per second).

Not only do pulsars spin, but they do so with such consistency that they can be compared with the extremely accurate atomic clocks that scientists use to measure time here on Earth.

Chapter 4. The beasts in the sky…

Quasars

In 1963 an unusual faint object was discovered which was also radiating a radio signal. Analysis showed it to be very distant, about 2 billion light years away from us. It was so far away that it must be radiating enormous power for us to see it and to receive the radio signals. The object was called a QUASAR which is an acronym for a 'quasi-stellar radio source' – *quasi* means *apparently* or *seemingly*, and *stellar* means relating to a star or star-like. So, the name means an apparently star-like object which is radiating a radio signal (and visible light since it is visible to us). Later quasars were found that were even more distant, up to 12.5 billion light years away. Remember that the Universe is understood to have formed some 13.8 billion years ago and the first stars were born about 400 million years later. So, the quasars that we were seeing were shining when the Universe was a child, only about a billion years old. But what object could be generating such continuous power that it is visible to us over these unbelievably vast distances?

We now understand that the object may look star-like, but it is in fact a galaxy which is releasing the energy from its centre [Fig. 4.2, above]. Such radiation is said to come from an active galaxy, that is, from a galaxy with an active centre or nucleus. The centre of an active galaxy is called an *active galactic nucleus* or *AGN* for short. The AGN is the powerhouse of the galaxy. And we know that the only object that that can generate such power is a supermassive black hole. Thus, a QUASAR is a young galaxy with an enormous active black hole at its centre that is shooting out jets of energy and matter which is travelling at almost the speed of light.

We saw earlier that as material falls into a black hole, it does so from an accretion disk of dust and gas that is spiralling onto the black hole. The enormous gravity of the black hole accelerates the material to near the speed of light. The tremendous friction that results causes the material to emit the unbelievable amounts of energy in the beams shooting out from the 'poles' of the black hole which we detect over the vast distances across the Universe.

The 1963 supermassive black hole has been calculated to have been generating more than 100 billion times the power of our Sun! I have deliberately said that the black hole would "have been generating" this power, since, while it was active billions of years ago, it may no longer still be active today. We have no means of knowing.

All that we have to go on is the light that left the galaxy all those billions of years ago. The black hole may well have switched off by now having used up its fuel of gas and dust that was orbiting around it in an accretion disk and generating the jets of matter and energy as it fell into the black hole. Interestingly, we do not see many active galaxies in the nearby Universe. This may mean that only younger active galaxies produce the jets which eventually get switched off (till new fuel is fed to the black hole), or it may mean that the active galaxies are a totally separate class of galaxies. We do not know for sure.

The energy from the jets is not necessarily constant. The emissions from different quasars can vary in strength and their spectra can show different characteristics such as broad and narrow spectral lines. As far as observation from the Earth is concerned, the characteristics we can see also depend on the direction in which the jets are facing. An interesting situation arises when the jets are facing directly at us, and we are looking down the throat of the quasar. Such a quasar is called a *blazar*. Blazars are strong radio sources, which also vary in strength over period of days. There are further sub-divisions amongst the quasars. But we will not go into these in this book.

Gamma ray bursts

The centres of active galaxies (AGNs) produce enormous power continuously. But let me now tell you about the most violent *explosions* known in the Universe which produce the brightest energy beams ever, but only for a short time. These are called gamma ray bursts (or GRBs for short).

Gamma rays are the extremely high energy radiation photons which are created when extreme energy-producing events occur. We met them when we discussed the Big Bang and the nuclear reactions that power the stars. We saw that the stars continuously convert the hydrogen nuclei in their cores into helium in a nuclear reaction which generates high energy gamma ray photons. But the explosion events we are talking about here release *bursts* of an unbelievable amount of energy in the form of gamma rays. It is because of such energy bursts that the GRBs get their name.

So much energy is released during a gamma ray burst that the event is observable, like the quasars, across billions of light years of space.

There are two types of gamma ray bursts. The long gamma ray bursts (long-GRBs) can last up to 500 seconds (about 8 minutes or so). The short gamma ray bursts (short-GRBs) typically last less than 2 seconds. Let us now see what causes these bursts, and how they came to be discovered, which is an interesting story.

We have spoken earlier about many of the extreme objects we find in space including white dwarfs, neutron stars and black holes. We have seen how these form through the death of stars. The larger stars when they die do so with a bang in a supernova explosion which leaves behind a neutron star or a black hole, the smaller ones die with more of a whimper which results in a white dwarf. But we know that even the white dwarf carries the potential for becoming a (Type Ia) supernova if it manages to collect additional mass so that it exceeds the Chandrasekhar limit.

Chapter 4. The beasts in the sky…

The long-GRBs are thought to be produced by the largest of supernovae, those due to collapse of supermassive stars which result in a black hole being produced. Another theory is that long-GRBs are produced when a white dwarf collides with a neutron star. The short GRBs are believed to occur when two neutron stars, two black holes or a neutron star and a black hole collide.

You will see from the uncertainty of the language I have used that gamma ray bursts are still the subject of active research.

If we could observe a GRB, it would be a cataclysmic event with an astonishing amount of energy generated. Together with the gamma rays, matter from the explosion would be ejected at near light speeds in two narrow beams along the rotation axes, which look like the jets from a quasar or a pulsar [Fig. 4.2, above].

We have seen that at the best of time a short-GRB can last only about 2 seconds and even the long-GRB usually no more than a few minutes. These bursts can occur without warning, anywhere in the Universe. So how do we see them, let alone study them?

But the first question to ask is: how did we find out about them at all?

GRBs were only discovered in 1969, and then only by chance, by a US military Vela satellite. That was the time of the cold war between the West and the Soviet Union. A nuclear ban treaty had been signed and there were many satellites in space to ensure that the ban was being obeyed and that no illegal nuclear tests were being conducted. Thus, the satellites were mainly focussed on seeing what was happening on Earth, not outer space. As nuclear explosions also generate gamma rays, the satellites were focussed on the Earth to identify any unexplained gamma rays that were found to be coming from the ground.

In 1969 the military satellites were picking up gamma ray signs which were causing concern. But to the consternation of the authorities, these signals were coming from behind the satellites, from outer space, and not from Earth. More satellites were launched, and further intensive study followed. This was difficult because no one knew where the next burst was going to come from, and when a burst was found there was an extremely short time to study it. Eventually a specialised satellite called *Swift* was launched in 2004 by NASA with international collaboration, specifically to study GRBs. In 2008 an even more advanced satellite was launched called *GLAST* (Gamma-ray Large Area Space Telescope). These telescopes not only had the capability of very rapidly swivelling around to face a GRB as soon as it is identified by the instruments on board, but they could also swiftly inform a network of ground-based telescopes, which in their turn could be brought to focus on the GRB for a longer and more detailed study.

The link between a GRB and a supernova was first established in 1998, and then confirmed in 2003 when physical evidence was found associating a GRB with a supernova at the same location.

When stars collide

We saw in the previous chapter that stars can form in a binary system. Such systems are quite common. Consider two stars in a binary system which are orbiting close to each other. Now imagine that one of the stars is much more massive than the other. What do you think will happen? If the 'heavier' star is near enough to the other star, it may steal gas from the other star's outer envelope. We will then have a streamer of gas flowing from one star to the other. As the stars are in orbit around each other, this gas will form a saucer like accretion disk around the more massive star as it falls onto it, in a similar way to an accretion disk around a black hole.

The more massive star will gain the mass which the other star loses. Now imagine that the more massive star is in fact a black hole [Fig. 4.4]. Slowly the gas will be stripped off the less massive star and start on its journey to oblivion to the black hole. It will spin around the black hole in an accretion disk and speed up as it falls, faster and faster. The gas particles will glow due to friction and emit higher and higher energy photons as they fall into the event horizon. The star losing its mass may be much larger than the black hole, but if they are near enough to each other, the black hole's gravity will always win. Slowly the star will be stripped of its gas, till its core glows naked. Even that may not be safe, and it may come to a sorry end, gobbled up by the insatiable monster next door.

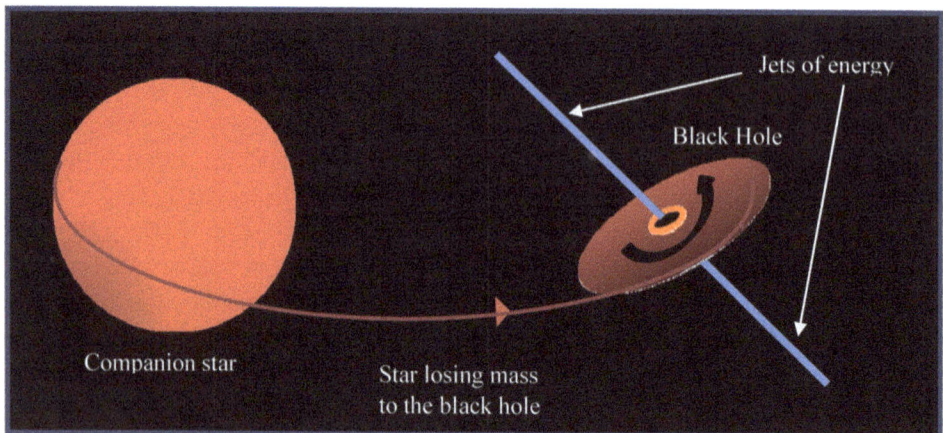

Fig. 4.4 Star and Black Hole in a binary (2-object) system.
The feeding Black Hole is stripping the star of its mass.

Chapter 4. The beasts in the sky…

Now assume another scenario in which the two stars in the binary embrace consist of a dying supergiant and a star in its middle age. The supergiant may well explode in a supernova leaving behind a white dwarf. We have seen that if the white dwarf is less than 1.4 solar masses, it can be quite stable. However, over time the other star may in turn become a supergiant and its gas envelope may reach to the white dwarf. The white dwarf can now steal gas from its much, much larger, but less dense 'twin' and so gain mass.

When the white dwarf's mass increases beyond the Chandrasekhar limit of 1.4 solar masses, its electron degeneracy pressure will fail to maintain the star's stability and it will collapse and destroy itself in a supernova explosion. Such a supernova caused by a white dwarf gaining mass from a neighbouring star is called a Type Ia supernova, as opposed to a Type II supernova which occurs when a star runs out of its fuel, then collapses and explodes.

If the white dwarf is much less massive than 1.4 solar masses, it may gain mass slowly. The hydrogen from the companion star accumulates on its surface and is compressed by the white dwarf's huge gravity. This gas can ignite in a runaway nuclear reaction, but it does so much less catastrophically than in a Type Ia supernova. Excess mass is thrown off before the white dwarf destroys itself, and the star lives to explode again another day. Such an explosion is called a nova; white dwarfs can trigger novae repeatedly.

An interesting thing about a Type Ia supernova is that we know the mass of the exploding star since this is 1.4 solar masses. Scientists are therefore able to work out the brightness of such an explosion. Thus, we know that every Type Ia supernova will have the same brightness and can serve as a standard candle. As we saw at the end of the last chapter, this knowledge provides us with an immensely powerful way of measuring distances to stars and galaxies.

A possibility is that the supergiant was large enough to leave behind a neutron star after its supernova explosion. When the above scenario of the remnant stealing gas from the nearby star now plays out, the neutron star can gain enough new mass to exceed its neutron degeneracy limit and to collapse spectacularly into a black hole.

A further possibility that can occur when two stars are in close orbit around each other is that they spiral towards each other and collide. If the collision is between two neutron stars or between a neutron star and black hole it can produce a short gamma ray burst leaving behind a black hole, a larger one in the second case.

When galaxies collide

When the Universe was young, stars came together drawn in by their gravity. They orbited around their common centre of gravity. Pretty soon the colony of stars grew till they were in their many millions in a small galaxy. Many other galaxies were forming. These galaxies then started coming together, the smaller ones being attracted to the larger ones. Thus, the hundred-billion-star galaxies were formed. This is the situation today.

If we look out into space, we see the stars in the Milky Way which is our galaxy. We can also see other galaxies. Some are relatively near to us; others are further away. The nearby galaxies are small in comparison to the Milky Way and are called *dwarf galaxies*. Two of the nearest galaxies are the Small Magellanic Cloud and the Large Magellanic Cloud. These can be seen in the Southern Hemisphere skies.

They were given their names because they looked like clouds obscuring parts of the night sky, the clouds being dust in the galaxies. There are other larger galaxies further away from us. The nearest large galaxy to us is the Andromeda galaxy (also known as M31) which is about 2.5 million light years away. This means, of course, that we are seeing the Andromeda galaxy as it was 2.5 million years ago [Fig. 1.11 in Chapter 1].

The galaxies in the Universe tend to be clustered together in groups, with large empty spaces between the groups. Our group of galaxies is called, reasonably enough, the *Local Group*. The Local Group spans about 4.5 million light years.

We know that the Universe is expanding because of the Big Bang. We therefore expect that all galaxies are moving away from each other. In general, that is so. But the galaxies in groups are close enough to be affected by each other's gravities. Thus, galaxies in the groups are attracted to each other and are in motion relative to each other. Some are moving in the same direction some in different directions. Their movement is being controlled by the gravitational attraction between the galaxies. It so happens that the Andromeda galaxy is heading straight for the Milky Way at some 300,000 miles per hour (480,000 kilometres per hour). It is expected that the two galaxies will collide in some 4 billion years or so in the future. What will happen then?

Surprisingly, there are virtually no collisions between the stars of two colliding galaxies. There is just too much empty space between the stars compared to their size. So, the stars of the two galaxies will generally pass by without crashing into each other. What happens to the gas and dust of the two galaxies is a rather different matter. The gas and dust particles collide and push and shove. They shine and glow due to the collision and friction energy they generate. Clouds that had been sedately floating in their respective galaxies can collapse, pushed by the incoming invader clouds. The process starts an extensive spell of star formation. If we look at two colliding galaxies through a telescope, we can see the blue glowing areas in their gas clouds heralding the new stars being born.

At the same time, gravity has a heyday. The stars are pulled hither and thither. The two galaxies can form enormous tails as they shoot past each other then come closer in a celestial dance. Eventually over hundreds of millions of years, they settle down into a new galaxy, a much larger one, obviously, than either galaxy was originally, but with an altered shape. Regularly shaped galaxies can become distorted. Spiral galaxies can lose their arms and end up combined as an elliptical galaxy, looking rather like a rugby ball.

Fig. 4.5 Colliding galaxies (NGC 2623/ Arp 243) – A rose made of galaxies.

When galaxies collide, we get a rather beautiful cosmic dance.

Galaxies moving towards each other may not actually collide but pass close by. In such a case, their gravitational attraction can disrupt the shape of one or both galaxies and start the burst of star formation. If one galaxy is smaller than the other, it is likely to be more affected. If the galaxies collide, one may pass right through the other. Once again, there will be disruption to the shape and star formation will occur.

When we look up into the sky using telescopes, we see many, many examples of colliding galaxies, of various shapes and sizes. The internet is full of dramatic photos. Just search for *colliding galaxies images* and you will be amazed by those that turn up. Many of the fascinating ones have been taken using the Hubble telescope.

How do we know this?

We can mathematically model a galaxy. In other words, knowing the size, mass, shape, speed, motion, and other characteristics of each galaxy, we can use mathematical equations to describe them and how they are moving, to see what would happen in a collision. We can use computers to display the models in graphical forms. Then we can run the model (rather faster than the galaxies take to evolve!) to see what happens. This is called a *simulation*. Finally, we can compare the computer simulation results with what we can observe in the heavens.

If astronomers find that the computer model is not correctly displaying what they see through the telescope, the model is altered and then re-run till real-life is correctly replicated. Once the model is accurately representing real-life, it can be used to study the subject more deeply. This is a good example of one of the ways that computers are used in scientific work.

You can see simulations of galaxies colliding on the internet.

Chapter 5

Our neighbourhood...

the Solar System

So far, we have been considering the Universe as a whole. We learnt how it was born in a Big Bang, and when. We found out that the Universe contains stars and galaxies and learnt how they came about and what makes them shine. We read about the many different types of stars that are in the Universe and how they are born and evolve and die. We have seen the many extreme objects that live in space and the enormously powerful events that occur involving black holes, neutron stars, supernovae, gamma ray bursts and galaxies.

Now I want to talk about something nearer to us: our home the Earth, and our neighbours the planets, asteroids, and comets which together with the Sun make up our Solar System. We will see how the Solar System was born, how big it is, what is in it, and what the future holds for it and for us.

How our Solar System formed

The Universe is huge. Within it are the stars and galaxies and dark matter. There are also clouds of gas and dust. Some of the material in the clouds is left over from the formation of stars, and some results from the subsequent death of stars - either through being spread as planetary nebulae or as debris from supernovae explosions. The gas is almost all made up of hydrogen and helium, but with traces of other gases as well. The dust is a collection of particles of different sizes that have condensed in space and clumped together due to gravity or electrostatic forces.

The electrostatic force is caused by static electricity. It is the force between objects with a positive or negative charge; you will remember that like charges repel and unlike charges attract. This force is the same one you may have experienced when you walked on a carpet and then touched a metal object seeing a tiny spark jump across giving you a small shock, or when combing your hair, you found that your hair-ends were attracted to the comb, or possibly when performing a party trick where you rubbed a balloon against a woollen jumper and then could stick it to the ceiling. The cause of all these effects is that the friction due to a rubbing action dislodges electrons (for example in a balloon) which then attract positively charged particles in the other object (such as the ceiling) and stick together.

The solid material in the clouds can be carbon and other molecules which were formed in the burning cores of the stars and expelled on their deaths. Some heavier elements are also present. A little bit of it is made of more complex molecules and may include organic compounds.

The gas and dust clouds are found floating within the galaxies and in the spaces between the galaxies. The clouds can be seen in the night sky as dark patches against the bright stars. Fig. 5.1 shows the famous "Pillars of Creation" with its 4 light year high cloud pillars in the M16 Eagle nebula where new stars are being born. Out in space the clouds generally are very cold, being only 10-20 K drifting far from any star. On the other hand, the temperature can be incredibly hot near the new stars being formed. Its temperature can therefore vary from a few degrees K to many million degrees K.

The clouds are extremely thin. If you collected the material from an Earth-sized cloud it would weigh barely a few kilograms. However, the clouds can be huge, and the material contained within them is often dense enough to stop light from stars behind passing through. This is what makes them look dark against the shining stars.

Fig. 5.1 The "Pillars of Creation" in the M16 Eagle nebula.

The dust also exists throughout space. In our solar system, every time a comet appears, it leaves behind a trail of dust. Each year, the Earth, in its orbit around the Sun, passes through the dust of many such comet trails. The dust particles the Earth meets generally burn up high in our atmosphere leaving bright streaks in the sky as they do. I am sure you would have seen these – we call them shooting stars or meteors (meteorites are objects large enough to fall to Earth without being burnt up completely). But there is also a lot of space dust that does not burn up. It is estimated that each day some 60 tons of this space dust falls on the surface of the Earth. Scientists have found space-dust in material that they collected off the roofs of houses.

We shall now see how one such cloud of gas and dust floating peacefully in space went on to form our Sun and then the Solar System [Fig. 5.2].

Sun

Occasionally, an event occurs that upset the serene existence of a cloud and gives it a shove that starts it moving. This could be a large body such as a star coming close enough for its gravity to affect the cloud, or perhaps, the shock wave from a nearby supernova explosion giving a push to the cloud while also (importantly) mixing the debris from the explosion with the material in the cloud. If the impulse given to the cloud is strong enough it could cause it to start collapsing towards its centre. The cloud's own gravity could itself be enough to cause it to collapse.

As a cloud collapses, it gets smaller and denser. The force of gravity pulling the cloud towards its centre increases, its speed of collapse increases, and its centre gets hotter. Eventually, as we saw in our chapter on Stars, the collapsing gas cloud gets dense enough and its centre gets hot enough for one or more stars to start to form, and eventually to shine as nuclear reactions start in their centres.

Fig. 5.2 Formation of the Solar System: Birth of Worlds.
Artist's conception of the dust and gas surrounding a newly forming planetary system.

This is how our Sun was born about 4.5 billion years ago. Remember that the Universe is almost 13.8 billion years old. So, the Sun contains the material created in possibly three or more earlier generations of stars. The Sun contains at least 67 elements. If you measure them, you will find that the main elements are hydrogen and helium with a little oxygen and carbon. There are also tiny amounts of nitrogen, silicon, and many others. Of these, only hydrogen and helium were formed in the Big Bang. All the other elements resulted from the nuclear reactions that caused the Sun to shine or were made in the centres of stars of earlier generations.

The Sun is big. Its radius is about 695,500 km (435,000 miles). Yet, in space terms it is just an average-sized star. Compared with some of the giants, it is a tiddler [Fig. 5.3]. However, compared with the planets that formed around it its size is impressive [Fig. 5.7]. For instance, the Earth's radius is some 6,370 km (3,980 miles). Thus, the Sun's radius is more than 100 times bigger than that of the Earth. This means that its volume is more than 1,000,000 (a million) times bigger than the Earth's, and so more than one million Earths would fit comfortably within it.

Chapter 5. Our neighbourhood…

The gas cloud was spinning in a flat disk around the forming Sun [Fig. 5.2]. So was the Sun itself at the centre. It is still doing so, rotating about its axis about once every 27 Earth days at its equator (it spins once every 35 days near its poles). The Earth we know rotates once every Earth day (or 24 hours). So, we rotate about 27 times faster than the Sun. Also, since the Earth is solid, as opposed to the gaseous Sun, it spins at the same rate all over.

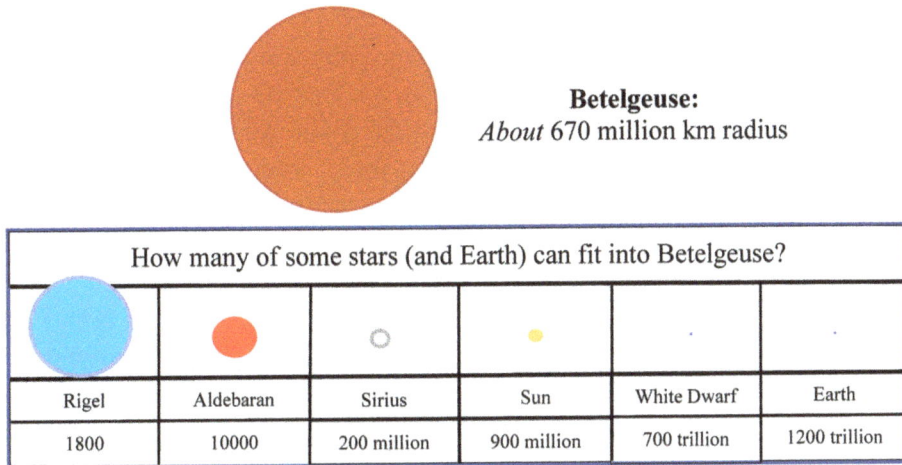

Betelgeuse:
About 670 million km radius

How many of some stars (and Earth) can fit into Betelgeuse?					
Rigel	Aldebaran	Sirius	Sun	White Dwarf	Earth
1800	10000	200 million	900 million	700 trillion	1200 trillion

Fig. 5.3 The Sun is a very average star.

The diagram shows the sizes of the Sun and some other stars, and
Earth, in comparison with one of the real supergiants of the Solar System.
Betelgeuse is a variable star – it expands and shrinks over time.
Its size given above is therefore approximate.

As the Sun shines it burns its main fuel, hydrogen, at a tremendous rate: about 600 million tons of it *every second*. It gets through more hydrogen than the mass of the whole of the Earth in about 70,000 years. It was born with more than enough hydrogen to last for 10 billion years. It has now been shining steadily for about 4.5 billion years and so is approximately halfway through its life.

The gravitational force of the huge mass of gas in the Sun results in enormous pressures in its central core, about 245 billion times higher than those of our atmosphere, and temperatures higher than 15,000,000 C. These enormous pressures and temperatures crush the hydrogen atoms together to produce helium and release energy (you should remember from the Chapter 3 on stars that this is called *hydrogen burning*). [Note: As Appendix B shows, the tunnelling effect of Quantum Mechanics is also essential for the conversion of hydrogen to helium].

Slowly (it takes about a two hundred thousand or more years), the energy from each nuclear reaction radiates outwards to the top of the Sun and out into space.

The layer at the 'surface' of the Sun which radiates the light is known as the *photosphere* and has a temperature of 5,500 C. The layer directly above this is called the *chromosphere* and is a little cooler at 4,300 C. Beyond the chromosphere a strange phenomenon occurs in a region called the Sun's *corona*: the temperature dramatically rises to more than 2,000,000 C (two million C). We do not yet fully know what causes the corona to be so hot.

The corona (meaning 'crown') is thought to extend some 8,000,000 km (eight million km) from the Sun [Fig. 5.4 (a)]. It is composed of plasma (positively and negatively charged particles) which streams away from the Sun in all directions. If you have been lucky enough to see a total solar eclipse when the Moon completely covers the Sun's disk, you may have seen the ghostly light of the corona surrounding the black disk like a crown. Even if you have not seen an eclipse, there are plenty of photographs that you can find online.

[An important warning: you must *never ever* look directly at the Sun. This can seriously damage your eyesight].

If you look at a close-up image of the Sun, you will find that its 'surface' is not smooth but full of what look like bubbles [Fig. 5.4 (b)]. The Sun does not have a surface as such since it is totally made of gas. There is a continuous churning of this gas as it rises from the interior, cools at the top, sinks, and rises again, just like water boiling in a pan.

Fig. 5.4 (a) Solar Eclipse 2017.

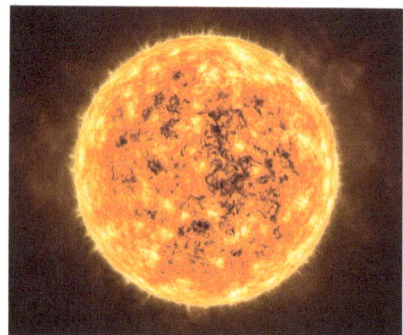

Fig. 5.4 (b) Closeup of the Sun (3D rendering).

In some images of the Sun, such as Fig. 5.5, you will see several black spots on the surface. These are called sunspots. The largest one in the image is almost 80,000 miles across, ten Earths could be laid across it. Sunspots are not black but appear so in photos since they are cooler (at 4,000 C or so) than their surrounding regions.

Sunspots occur in pairs. They are regions where the Sun's magnetic field breaks out of the Sun, and where it re-enters. As the magnetic field loops from one sunspot to the other one of the pair, it carries tremendous magnetic energy. Occasionally, the field snaps which leads to huge eruptions of the Sun's material in the form of solar flares or storms called coronal mass ejections [Fig. 5.6]. If these eruptions head towards the Earth, they can damage our satellites, result in magnetic storms on Earth and give impressive displays of aurora over our polar regions. One such storm in 1989 resulted in the black-out of much of the city of Toronto in Canada.

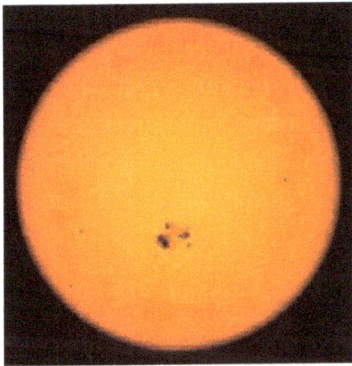

Fig. 5.5 Sunspots on the Sun.

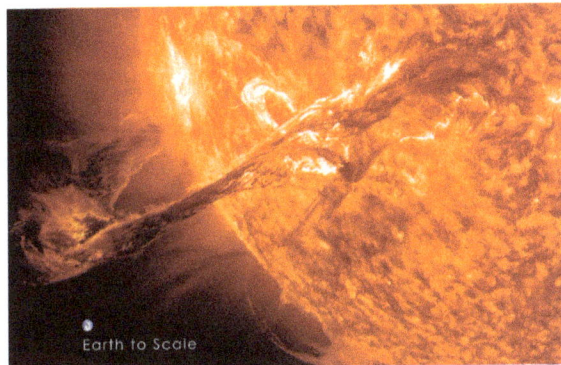

Fig. 5.6 Solar flare and mass ejection from the Sun.

The frequency of the sunspots on the Sun goes through an 11-year maxima/minima/maxima cycle. Each time the sunspot count goes through its 11-year cycle, the Sun's own magnetic field reverses direction giving an overall 22-year cycle.

You can see that the Sun is a complex object. But we must remember that remarkable as it may seem, the Sun, big as it is and as hot as it is, is just an average-sized star shining with an average brightness. We have seen that there are many bigger stars in the Universe, even supergiants which can be 50 or more solar masses. In the universal scheme of things there is nothing particularly special about the Sun, or even about our galaxy the Milky Way.

Yet the Sun is not any old star. It is *our* star. It is the only one we have, without the Sun neither our solar system nor we would exist. There would be no light, no warmth, just a cold cloud of gas and dust floating in space waiting for a nearby supernova to give it a shove.

The formation of the Sun was just the start for our Solar System. There are many more things before the story of the building of our Solar System is complete. There are still many more fascinating things in our neighbourhood to learn about. Let us now turn our attention to what else was happening in the cloud apart from the formation of the Sun at the centre.

The planets

When the nuclear reactions started in the Sun, it was still surrounded by its dense dust and gas cloud. The cloud was in the form of a flattened disk which was spinning around the glowing Sun. We saw in the earlier chapter that the spin is a natural result when a fluid or gas falls towards its centre. In the spinning disk, the dust slowly started to combine into larger and larger clumps as the initial particles came together due to the electrostatic forces between them (we have discussed the electrostatic force earlier in the chapter). Their gravity at this stage was too small to be the main attracting force.

The clumps attracted other nearby particles and clumps and grew into larger rocks. As the rocks grew bigger, their gravity also grew till they were able to attract still more rocks due to this force. There were a lot of collisions between the growing rocky bodies; sometimes the colliding rocks shattered into smaller pieces and then came together all over again, sometimes they stayed stuck together and grew even larger. As the rocks grew, so did their gravity, and even more of the nearby orbiting matter coalesced onto the growing body.

The collisions became ever more violent, and the rocks heated up and melted. They had started to spin as the Sun had done earlier. The largest pieces took on a spherical shape (at about 600 km diameter) as their own gravity grew big enough to pull their composite material to the body's centre.

Over time, these spherical rocks grew bigger still into a planetoid and then into a planet orbiting the Sun. It is believed that the rocky planets took several million years to develop, but the gas giant formed in under a million years.

The planet formation was taking place simultaneously in many different parts of the cloud. The planets that were forming near to the centre of the cloud were nearer to the Sun and so were in an area that was much hotter than the far reaches of the cloud.

Stars when they form develop a 'wind' of charged particles which blows away from them. The Sun was no exception. The wind generated by the newly formed Sun was blowing away the lighter gases in the cloud to the colder far regions of the newly forming solar system. The heavier solidified matter was, therefore, mainly to be found near to the Sun. This resulted in the planets that formed near to the Sun generally being rocky.

Fig. 5.7 shows the sizes of the planets compared to the Sun. We will discuss the planets in the sequence in which they appear in the Solar System as shown in the diagram. The rocky ones come first, before the giants of the Solar System, the distant gaseous planets. Finally, we will look at the dwarf planets (not shown in the diagram), which are too small to be considered amongst the eight major ones.

Chapter 5. Our neighbourhood…

The terrestrial planets

Four rocky (also called *terrestrial*, meaning Earth-like) planets orbit around our Sun: the nearest one to the Sun is Mercury, next is Venus followed by Earth, and finally Mars.

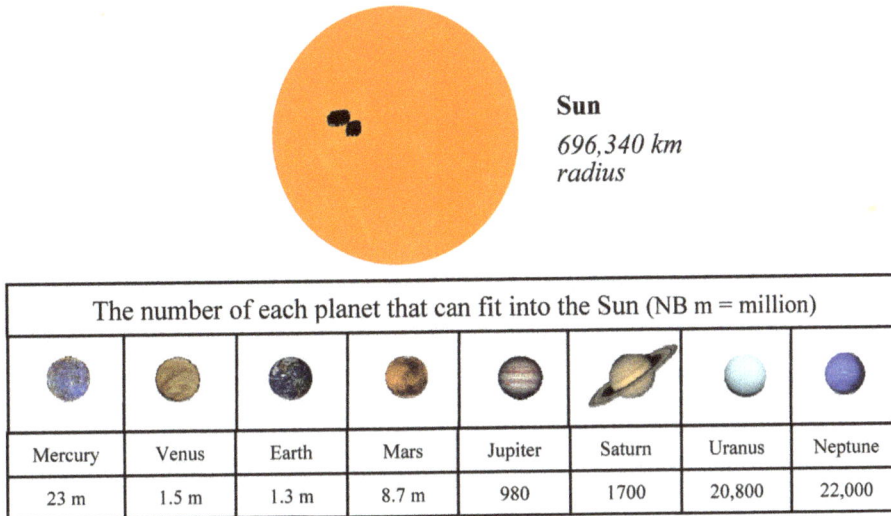

Sun
696,340 km radius

The number of each planet that can fit into the Sun (NB m = million)							
Mercury	Venus	Earth	Mars	Jupiter	Saturn	Uranus	Neptune
23 m	1.5 m	1.3 m	8.7 m	980	1700	20,800	22,000

Fig. 5.7 Sizes of the planets compared to the Sun.
The planets (not to scale) are shown in order of their orbits around the Sun.

All the terrestrial planets have the same basic structure. I will describe this briefly. The details are less important than your understanding the general format, so you can appreciate the differences that exist between the planets.

The main point about the rocky planets is that they are generally solid. But if you were to drill right down to the centre you would find that they are not quite all the same. I will show the Earth's structure here and point out the specifics of the other planets as we discuss them [Fig. 5.8].

There is a solid crust at the top of all rocky planets in our Solar System. Below Earth's crust is a layer called the mantle, followed by the core. The mantle is divided into the upper mantle and the lower mantle. The upper mantle is hotter and weaker than the crust and can flow (very slowly). The lower mantle is under more pressure and is more rigid than the upper mantle. The core is also divided into an outer core and a solid inner core. The Earth's core is mainly iron with some nickel, the outer core being fluid and the inner core solid. It is the flow of material in the outer core that is responsible for the generation of the Earth's magnetic field.

While all our terrestrial planets fit this general pattern there are differences, such as in the sizes and presence of the layers and whether there is flow of material. These differences are crucial in defining the nature of the planets, and in determining whether the planet is considered 'dead' or 'alive'. As I am sure you can guess, the most complex structure is the Earth's.

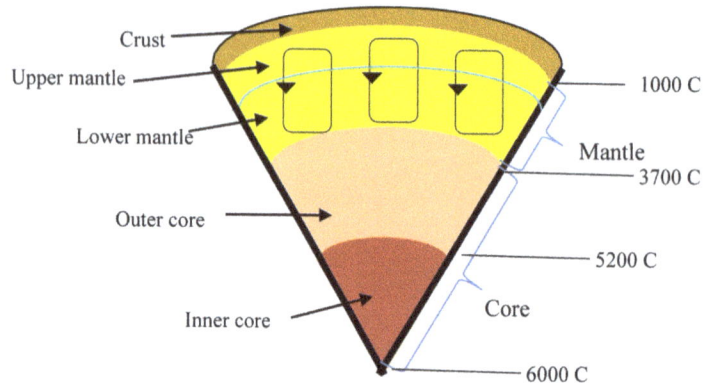

Fig. 5.8 Section through the Earth.
The components vary from planet to planet.
(Not to scale)

Apart from their internal structure, the biggest difference between the planets is their location in relation to the Sun. We will consider this as we review the planets. We will measure this distance in kilometres (km), but also in astronomical units (AU). The astronomical units compare the distances using the distance between the Sun and Earth (150 million km) as 1AU.

Let us look at the planets in sequence from the nearest to the Sun to the farthest. We will start with the terrestrial planets, specifically Mercury, then Venus, Earth, and Mars in turn [Fig. 5.9].

Fig. 5.9 The terrestrial planets.
Mercury, Venus, Earth, and Mars are shown in approximate relative size.

Chapter 5. Our neighbourhood…

Mercury

Mercury [Fig. 5.9, terrestrial planets] is the nearest planet to the Sun. From its surface, the Sun would appear three-times as large as it does to us on Earth. It has a peculiar egg-shaped orbit which ranges from 47 to 70 million km from the Sun, and travels at 47 km per second around it. This shape of the orbit results in a strange phenomenon: as seen from some parts of the planet, the morning Sun rises, sets, and rises again, and does the same thing in reverse at night.

Mercury rotates three times for every two times it goes around the Sun. What this means is that a year on the planet (which is the time it takes to go around the Sun) is one and a half (1.5) of its days (the time it takes to rotate). If Earth rotated only 1.5 times per year, our day would be 243 of our current days long and we would have a birthday every 1.5 days.

Because Mercury is spinning so slowly, the side facing the Sun during its day gets baked by the Sun to a crispy 430 C (over 700 K), while the side that is facing away from the Sun is frozen solid at -180 C (about 90 K). Should astronauts visit Mercury's surface they would need highly efficient space suits to cope with these temperature extremes.

Mercury is also the smallest of our eight major planets having a radius of only 2,440 km just a little bigger than our Moon (radius about 1,750 km). It orbits the Sun in only 88 Earth days, and spins on its axis every 59 Earth days. Its surface is the most heavily cratered of any object in the solar system: much more than that of the Moon. This means that in the distant past Mercury's surface was bombarded by asteroids and comets, just like the Moon's and Earth's surfaces. There are also plains, again like on the Moon. This suggests that there were ancient lava flows. Today, however, there is no volcanic activity.

Most of Mercury's interior volume is taken up by its large metallic core 2074 km in radius, which is probably partially molten giving it a (very weak) magnetic field. Its combined crust and mantle outer shell is only some 400 km thick.

Mercury is an inner planet (one whose orbit lies within the Earth's orbit). Since Mercury is so near the Sun, it is also near to the Sun in the sky for us on Earth which makes it difficult to spot even through a telescope. This is particularly so as it is also so small. However, if you have a clear view of the western horizon just as the Sun is setting when Mercury is an evening object, it is worth trying to see it just after the Sun disappears. When it is a morning object, you will need to rise early and look for it in the eastern sky just before the Sun rises.

Mercury was the Roman god of trade, merchants, travellers, and thieves. He was also the messenger of the gods. Mercury is shown in sculpture with wings on its sandals to depict speed. Mercury the planet lives up to this billing by moving rapidly in the sky for us from being a night object to a morning one.

Venus

Venus [shown in Fig. 5.9, terrestrial planets, without its cloud cover] is a spectacular sight in the night sky. It outshines every other star and planet and is only challenged by the Moon. As Mercury, Venus is also an inner planet and shows the similar characteristics of waxing and waning as a crescent, and changing from a night to an early morning object in the sky as it orbits around the Sun. In Venus's case, since it is further from the Sun it can climb higher in the sky than Mercury and is extremely easy to spot. Its crescent shape is easy to see even through a small telescope or good binoculars.

Venus was named after the Roman god of beauty. When you see it in the sky it is not too difficult to imagine why the name was given. People believed that it must be exceptionally beautiful on its surface too. They were to be bitterly disappointed as we shall shortly see. But first here are some facts.

Venus is 108 million km (0.73 AU) from the Sun. It is virtually a twin of the Earth in size, being 12,104 km in diameter compared to the Earth's diameter of 12,756 km and having a mass 81.5% that of the Earth. When seen from the Earth, even with the biggest telescopes, it is found to be completely covered with clouds and no part of its surface is visible. This is the reason Venus is so bright: the clouds reflect most of the light that falls on the planet. Scientists measure how efficiently an object reflects light by its albedo, which is the proportion of the light that is falling on an object that is reflected. The Moon has an albedo of only 0.12 (that is it reflects 12% of the light falling on it) which makes it dark. Earth's albedo is 0.3 reflecting 30% of the light it receives. Venus on the other hand has an albedo of 0.75. The object with the highest albedo in our Solar System is Enceladus, a moon of Saturn, which is covered by ice and has an albedo of 0.99.

Venus is 108,000,000 km from the Sun. It goes around the Sun in an orbit that is almost circular. If we measure Venus' orbit and rotation from Earth, we find that it takes 225 Earth days to complete each circuit of the Sun, and, interestingly, we find that it rotates about its axis once every 243 Earth days. Thus, Venus' day is longer than its year.

Like Earth, Venus has an iron core which is about 3,200 km in diameter. Above the core it has a mantle which churns, like Earth's mantle, giving it a weak magnetic field. Final there is a thin crust. Venus has many volcanoes created by its surface that moves as its mantle shifts.

We have noted that Venus like Mercury is one of the two inner planets in the Solar System, its orbit lying within the Earth's. Being an inner planet [Fig. 5.10] means two interesting things for both Mercury and Venus. One is that the planet has phases just like our Moon. If you see it through a telescope you will see that the planet's surface is a crescent which waxes and wanes, that is, it grows to become full then becomes smaller. If you think about it, this is simply because we are seeing the planet partly lit by the Sun, changing as it moves around in orbit. We see the planet's full face when it is on the far side of the Sun unless it is directly in line with the Sun of course. Inner planets are also called inferior planets and the outer ones superior.

The other interesting thing about Mercury and Venus is that as they move around the Sun the planets change from being evening objects in the sky for us setting after the Sun, to becoming morning objects rising before the Sun. Again, this is due to it being inner planets [Fig. 5.10]. Note: all planetary orbits are in the same anti-clockwise direction as seen from 'above' the Sun.

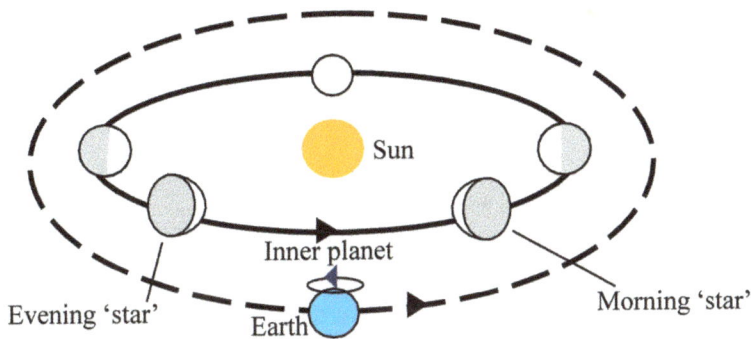

Fig. 5.10 Phases of an inner planet as seen from Earth.
The planet moves between being an evening and a morning object.

Venus is also unusual in that it rotates in the opposite direction to most of the other planets in the Solar System; this is called retrograde rotation. The only other planet that does so is Uranus. Venus must have been knocked upside down in a collision early in its life. For us on Earth the Sun rises in the East and sets in the West. On Venus it rises in the West and sets in the East.

If you were on Venus and started a clock when the Sun is directly overhead (your noon) to measure your day, you will find that by the next time the Sun is overhead (noon one day later) 117 Earth days have passed. This is called Venus's solar day.

The difference of 126 days between the 117 days as measured by you on Venus, and the 243 days as measured from Earth is due to the combined effect of our orbit around the Sun and the retrograde rotation of Venus.

We are not yet done with the strange facts about Venus.

For a long, long time nobody knew anything more about Venus than that it is a very bright object in the sky encased in white, shiny clouds. People had no idea what its surface was like and imagined a paradise and named it after the goddess of beauty.

Then in 1962 a US probe, Mariner 2, was sent on a mission to fly-by 22,000 miles above the planet. The probe confirmed the cloud cover. It also discovered that the atmosphere was made of carbon dioxide and measured the temperature and pressure on the planet, both of which were found to be amazingly high.

The pressure on the surface was found to be 90 times that on Earth, enough to squash any human, and its temperature was a blistering 475 C, which is more than hot enough to melt lead. Analysis of the clouds showed that they were made of carbon dioxide and sulphuric acid which is one of the most corrosive substances we know and highly dangerous to humans. Far from a paradise, Venus was near enough to being Hell.

Many other probes followed. In 1967 the USSR Venera 4 spacecraft managed to enter Venus's atmosphere and used a parachute to descend to 25 km. In 1972, Venera 7 soft landed on the surface and lasted just long enough to confirm the extremely high temperature and pressures. Other spacecraft discovered lightning and thunder on the surface and took colour photographs of the ground around the probe. From 1990-1994, the US Magellan craft mapped Venus' surface in detail using radar from orbit. This is how we have images of Venus without its cloud cover, as in Fig. 5.9. Then in 2006, the European Space Agency (ESA) sent a spacecraft, Venus Express, to orbit the planet and collect more data.

You will see that by now we know quite a lot about the planet. We have found that its surface is very Earth-like with mountains and valleys, and what look like riverbeds. It is however surrounded by a noxious atmosphere under enormous pressure and unbearable temperatures. How did it get this way?

A likely scenario is that Venus started very much like Earth, with flowing water and reasonable pressure and temperature. However, its atmosphere was primarily carbon dioxide, with some sulphur bearing gases. You would know that here on Earth the atmosphere is mainly a mixture of nitrogen and oxygen. We breathe in oxygen and breathe out carbon dioxide. Carbon dioxide is taken in by plants which in turn give out oxygen. So, carbon dioxide is very much part of the life cycle on Earth. Carbon dioxide however also has another characteristic. It is a greenhouse gas. In other words, it traps heat. Too much of it means that the surface temperatures rise. This is a big concern we have on Earth too; carbon dioxide levels are rising, resulting in global warming. Scientists are seriously concerned that we should control and reduce the amount of carbon dioxide in the atmosphere to stop the Earth getting too warm.

Chapter 5. Our neighbourhood...

On Venus, things got out of control and we had 'a 'runaway' greenhouse effect'; the temperatures kept rising and all the water on the planet boiled away. The water vapour in the atmosphere also acted like a greenhouse gas which made matters worse. The water combined with sulphur compounds in the atmosphere to form sulphuric acid. Huge clouds formed in the atmosphere further acting as a blanket to keep in the heat. The result was that the temperatures rocketed leaving the planet as we now find it - a dead place where life could not possibly exist.

But is this truly so? You will know that it gets colder the higher you go; the mountain tops are colder than the plains. It has been calculated that some 50 km above the surface within the Venusian clouds, the temperature could be down to 30-50 C which is tolerable to some life forms such as certain bacteria. So, is there life on Venus? Very unlikely, but possibly, is the conclusion. Space agencies have considered sending a probe to investigate.

Earth

The third rock from the Sun, as it has sometimes been called, is Earth [Fig. 5.9, terrestrial planets], the third planet counting away from the Sun. We have already said that it is about 150 million km (1 AU) from the Sun. It is our home planet, the only home we have (at least for now). So far, it is the only place in the whole Universe that we know has life. We have been busy looking for life elsewhere, but to-date have not found anything. But we search on.

We will learn about life on Earth, and in other places, in the next chapter. Here we will focus on how Earth came into existence, how its continents and seas formed, how its atmosphere came into being and whether it was always as it is now. We will discuss the dangers that face the Earth. The Earth itself will outlast most of the threats. However, as with all things, even it will die one day. But how will this happen and when? We will discover what the scientists think. We will find out if there is anything that we humans can do to save ourselves.

The Earth is one of the four terrestrial planets. We have talked about little Mercury and hot Venus. Earth is next, the third in the list. The last of the four is Mars which we will consider in the next section.

As was the case with all the terrestrial planets, Earth was born of the same cloud of gas and dust as the others. But it is different. It has continents and seas and an atmosphere; it is green and teeming with life. The main reason for the difference is the distance that the Earth is from the Sun. Its orbit is in what is known as the *Goldilocks zone*: not too hot and not too cold. Remember that the Sun's surface is at a blistering 5000 C. The further we go away from the Sun, the lower is the temperature. Something exceedingly important happens over a specific range of distances from the Sun. This range (or zone) stretches, in the case of the Sun today, from about 142 million km up to about 230 million km.

The reason why this range is so important is that the temperature within it enables water to stay liquid, provided there is enough atmospheric pressure. We know that liquid water is the one essential ingredient for life as we know it to exist. For obvious reasons, the Goldilocks zone is also sometimes called the *habitable zone* - the area of the solar system that is suitable for living things to inhabit.

The only planet that sits within this zone in our Solar System is the Earth, which at 150 million miles from the Sun is just within. Mars is on outer edge of the zone or just beyond. It has too little atmosphere and is too cold for liquid water to exist on its surface. Venus is too near the Sun and is simply too hot in any case due to its runaway greenhouse effect.

How the Earth (and the Moon) formed

We have seen how the planets formed in the disk of the gas and dust cloud rotating around the Sun by particles in the cloud colliding and coming together, first due to electrostatic forces then due to gravity. The Solar System went through the process some 4.5 billion years ago, not long after the Sun itself formed. As the gas cloud was collapsing to form the Sun at its centre, the planets were forming in parallel orbits within the rotating gas cloud.

The Earth too grew by accumulating material from the cloud into a ball of molten rock, heated by collisions with other rocks in its orbit. A crust would form on the planetoid as the molten rock cooled but would then be smashed up and melted again by the heat of new collisions. Gravity would bring the pieces of the smashed sphere and the colliding rock back together each time to form a larger sphere. As the sphere grew, the heavier material in it, such as iron, sank to the centre due to gravity. Thus, the composition of the Earth started being differentiated with the lighter material at the top and the heavier material near the centre. You can see that space within the cloud at that time was not a pleasant place to be.

One other feature of this cataclysmic stage in the Earth's formation was that there was no Moon. Let us see how the Earth completed its own formation and gained a Moon in the process, because the two processes are intricately connected.

There was a long argument amongst scientists as to how the Moon came to be. Some thought that it formed elsewhere in the Solar System and later wandered into the Earth's gravitational space and was captured by the Earth's gravity. Others thought that it formed at the same time and out of the same rock debris as the Earth. Still others believed that it split from the Earth soon after the Earth formed and pointed to the Pacific Ocean as the depression it left behind. But none provided a satisfactory answer.

The scientists then got hold of Moon rocks brought back by the astronauts from the Moon's surface. Analysis of this material gave them a good idea of the Moon's composition. They knew that the Moon did not have an iron core unlike the Earth (the density of the moon rocks was too low), that its rocks must have been extra heated (they did not contain water) and that the Moon and Earth were formed at the same distance from the Sun (by the chemical analysis of the rocks). The results were at odds with what the theories had been predicting. So, it was back to the drawing board. Then in the mid-1970s a new theory (The Giant Impactor Theory) was put forward. This said that another planet, about the size of Mars had formed in an orbit that would lead it to collide with the Earth.

This planet has been called Theia, named after the Greek Titan goddess of sight and heavenly light, who was also the daughter of Heaven and mother of the Sun, Moon and Dawn. About 50 million years after the Earth had formed, Earth and Theia collided. It was the biggest impact that Earth has ever suffered, many millions of times as powerful as the one that killed off the dinosaurs. Luckily, the collision was not head-on, else the Earth may well have been vaporised or shattered into tiny pieces. The collision is thought to have been a glancing blow which melted Earth and the other planet and threw the hot material from the upper parts of both planets out into space.

What was left of Theia then combined with the damaged Earth to form a new, larger Earth. The ejected material, which did not include material from the colliding planets' heavy iron core because of the glancing collision, went into orbit around the new Earth.

Slowly as time passed, gravity got to work; some of the material fell back to the surface of the Earth, the remainder collected in orbit around this new Earth to form a sphere which today we call the Moon.

Remember, this is a theory and is being questioned and refined as theories should be. But it is the best we have today, one that best fits the facts as we know them.

We shall refer to the Moon again later, but for now let us continue our story of the formation of the Earth.

The Earth continued to be bombarded by the space debris in its orbit and by asteroids and comets. We will read later in the chapter about how asteroids and comets were created and how they came to be in Earth's orbit. The bombardment of the Earth and other rocky planets resulted in surface craters, some huge and some small. The craters can still be seen today, most clearly on the Moon and Mercury where there has been virtually no erosion. There was lava on the Moon, early in its history, when it was still partly molten and had volcanic eruptions. This lava flooded out of cracks in the surface and filled many of the craters.

Earth also has a few craters remaining from the past impacts. Examples of clearly defined craters on Earth are the 215,000-year-old Manicouagan crater in Quebec, Canada, and the Meteor crater [Fig. 5.11] in Arizona, USA, which is about 50,000 years old. The largest crater on Earth is the 300 km diameter, 2 billion years old Vredefort crater in South Africa. Most of the older craters are no longer to be seen on Earth having been obliterated by erosion due to wind and rain, and by the renewal of the Earth's surface due to lava flows over the millions of years.

Fig. 5.11 Craters on Earth: Meteor crater, Arizona, USA.

Now back to the story of Earth's formation.

Once most of the marauding projectiles had cleared out of the planetary orbits, the terrestrial planets' crust slowly solidified. The rocks and comets which had been bombarding the newly forming planets contained many different minerals and other substances such as water. This material mixed with the molten rocks and sank into the interior of the planet. The heavier elements, such as iron, fell through to the centre. Material which was radioactive began to emit radiation which heated the surrounding interior rocks.

As the Earth cooled, a banded structure evolved in the planet with a of crust, mantle, and core [Fig. 5.8]. The lighter substances floated to the top, and the heavier material fell to the centre. The iron core was solid though under tremendous pressure from the whole mass of the Earth above. The layer above called the outer core was molten iron and churned around the inner core.

The fluid, flowing in the outer core is particularly important, as it is the cause of the Earth's magnetic field. Our magnetic field shields us from the worst effects of the damaging cosmic rays which are charged particles that come from the Sun and outer space. Some of these get through the protection of the magnetic field and are directed by the magnetic field towards our north and south poles. This results in the spectacular aurora borealis (northern lights) and aurora australis (southern lights) seen in the Arctic and Antarctic regions of Earth, respectively [Fig. 5.12].

Chapter 5. Our neighbourhood...

The aurorae occur when charged particles from the Sun interact with gas atoms in the upper atmosphere. The colour of the lights depends on which atoms are excited: oxygen can generate green or red, nitrogen red or blue. Often to the naked eye the colours are very faint or even missing and the aurora appear pale because they are too faint to be sensed by the eye's colour detecting cells. Aurorae appear on other planets as well, such as on Jupiter and Saturn.

Fig. 5.12 Aurora borealis – northern lights in Iceland.

Between the crust and outer core [Fig. 5.8], there are two regions called the upper and lower mantle. These regions also flow in slow currents through a process called convection. Convection occurs when liquid in a container is heated from below. The heated material rises to the top, cools, and then drops back to the bottom where it is heated again and once more rises to the top, and so on. Such a circulating current is called a convection current.

Convection currents can be easily demonstrated by using a saucepan of water on a hot ring or plate. A little food colouring will enhance the effect. Here the heat is provided by gas or electricity. Inside the Earth, it is provided by heat trapped during the Earth's formation and by the heat generated by the radioactive material within the Earth, as we mentioned above. Even though the mantle is solid the pressure and temperature are so high that the rocks can behave like liquid and flow, albeit extremely slowly.

We have seen that the churning liquid upper core generates our magnetic field. The currents in the Earth's upper and lower mantle give us plate tectonics. What is 'plate tectonics'? Well, it is a theory about something especially important for us. Let me describe it.

The Earth's crust is not a complete, solid cover over the planet; it is cracked into seven major pieces called *tectonic plates* [Fig. 5.13]. These plates carry two types of crust: the ocean crust and the continental crust. As the names imply the ocean crust carries the Earth's oceans and the continental crust forms the land. The ocean crust is denser, though thinner, than the continental crust. The continental crust therefore floats higher than the ocean crust. These plates move about on the top of the Earth, driven by the currents in the mantles. This means that the continents and the oceans on Earth are continuously in motion, colliding, retreating, and changing shape.

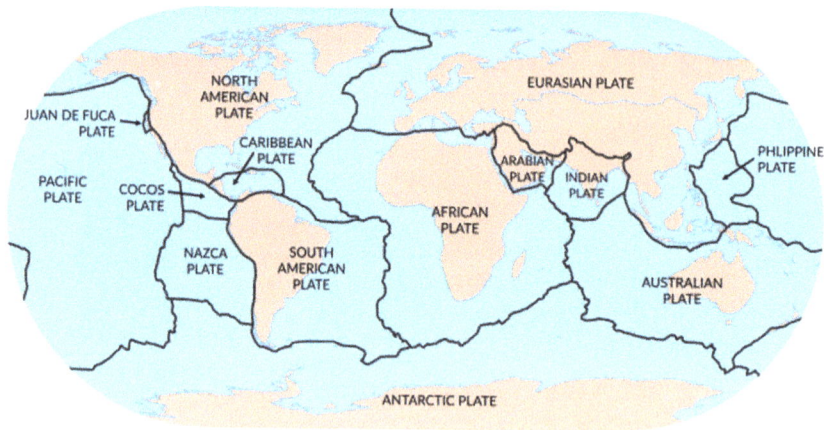

Fig. 5.13 Earth's major tectonic plates.

Today, Europe and Northern Asia are part of the Eurasian plate. Next to this plate, to the west, is the North American Plate. The divide between the two plates runs down the middle of the Atlantic Ocean. These two plates are moving apart – the ocean is getting wider and North America is moving further away from Europe at the rate of about 1 to 2 inches (2 to 5 cm) per year. This is about the rate that your nails grow.

Where is this extra material coming from? If you were to go down to the bottom of the Atlantic, you will find a long mountain range (ridge) running north-south down its middle under the sea. Here new material is rising from the mantle, deep beneath the sea crust and spreading on each side of the ridge to create a new ocean floor. If you follow the ridge to the north, you will come to an undersea mountain that rises above the sea level. Above the sea level, this mountain is a country called Iceland. The dividing ridge runs right through Iceland, and as it spreads Iceland is widening and new land is being created. You can travel to Iceland and see this for yourself.

Chapter 5. Our neighbourhood…

What is pushing the Atlantic Ocean apart? If you could see the mantle below the crust, you would find the part of the mantle that is on the European side is flowing to the east and the other part below the North American plate is moving west. So, the currents in the mantle are pushing the two plates apart and the oozing material is creating the new sea floor and land.

But surely, there is a problem. If new land is being created, why is Earth not getting bigger?

The reason is that elsewhere the same amount of material as is coming out to create new sea floor and land, is disappearing back into the mantle. The result is that the continents of America and Europe are moving apart, and the Atlantic is growing wider, but the other side of the continent, the North American plate is pushing against the Pacific plate. Remember I said that the continental crust is lighter than the ocean crust. So, when the North American continental plate hits against the Pacific Ocean plate, the continental plate floats on top of the Pacific plate which gets pushed underneath the crust. Similarly, on the west of the Pacific Ocean, the western side of the Eurasian plate crashes against the Pacific plate which is pushed underneath. The Pacific Ocean is thus getting squeezed from both sides and is shrinking, and the Eurasian plate in the Far East is getting closer to the North American west.

Of course, the plate being pushed back into the mantle does not go under easily. Huge tensions develop in the rocks making up the two crusts. Stresses build up until something gives; the rocks shatter resulting in earthquakes. You will therefore find that as you trace the boundaries of the plates and see where the plates are colliding, that that also defines the regions of the world where earthquakes occur. Such a boundary around the Pacific Ocean is called the Pacific ring of fire [Fig. 5.14].

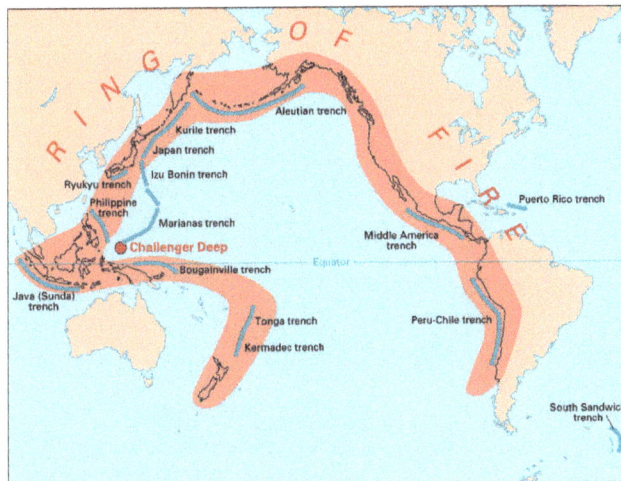

Fig. 5.14 The Pacific ring of fire.
The boundary of the Pacific Ocean where the plates are colliding.

Moreover, when one plate goes underneath another, the heat generated can melt the plate above [Fig. 5.15]. The molten material being lighter than the solid rock, rises through cracks in the crust on top to explode as a volcano. Thus, we find volcanoes near the areas where earthquakes can occur. An example of mountain ranges that formed due to the ocean plates flowing under continental plates are the Rockies and the Andes, which run virtually the whole western lengths of North and South America. These were formed by the Nazca and Pacific plates being pushed under the North and South American continental plates.

Fig. 5.15 When tectonic plates collide… mountains are formed.

When two similar plates (continental or oceanic) collide, then, since both are of equal density, they crumble and push up the crust into mountains at the boundary where they are colliding.

The collision of the African plate into the Eurasian plate resulted in the Alps. The collision of the Indian plate into the Eurasian plates produced the Himalayas. Interestingly, the collisions continue to this day, and the Himalayas and the Andes are growing higher even as we speak.

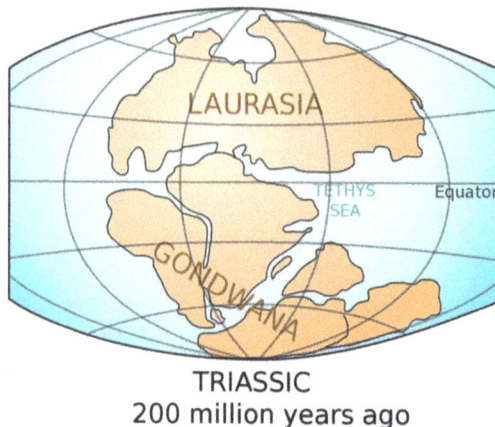

TRIASSIC
200 million years ago

Fig. 5.16 Earth 200 million years ago.

Chapter 5. Our neighbourhood…

The plate tectonic processes have continued over the ages, resulting in the continents moving about on the surface of the Earth. The map of the world is in continuous change although very, very slowly. If we look at the world map showing the land and seas during the Triassic period around 200 million years ago, we find that the continents had drifted together into two major land masses: Laurasia, which comprised of North America, Europe, and Asia, and Gondwana, which included Africa, South America, India, Australia, New Zealand, and Antarctica [Fig. 5.16]. At that time India had still to collide with the Eurasian plate and therefore the Himalayas did not exist, and neither did the Alps.

The most striking demonstration of this theory is to see how neatly the edges of the continents fit into each other, such as South America into Africa, Antarctica into Australia.

Slowly the continents drifted and over the next 200 million years came to be where there are now. I wonder where they will be after the next 200 million years.

Of course, as these lands drifted around, they had a major effect on how the sea currents and air currents moved. That then influenced the climate around the world. So, over the ages the climate of the world has kept changing. At one time the whole world became covered in ice (this is called the "Snowball Earth"), another time the climate became tropical. The changing climate affected the plants that could grow, and the animals that could survive. As you can see Earth has always been a very dynamic place.

We have spent a while looking at the land masses, the continents, of the Earth. But we know that there are other elements that exist: water in the oceans and rivers, and gases in the atmosphere. How did these come about?

Land and Water

We have seen that the Earth after it cooled formed a crust that resulted in plates that floated on the mantle. We also saw that the crust was of two types: a lighter crust, which is called the continental crust, and a denser crust which floated lower than the continental crust and which is called the ocean crust. Today the continental crust forms the continents, the ocean crust forms the bottom of the oceans and water fills the seas.

Water is so common on Earth that we take its existence for granted. But where did it come from?

As we have seen earlier, each water molecule is formed of 2 hydrogen atoms and 1 oxygen atom (written scientifically as H_2O). Hydrogen was formed in the big bang, and oxygen is one of the elements produced in a star's core. These chemicals came together in space and reacted together to form water. There is plenty of water in space. It exists mainly in the form of ice or vapour and in compounds formed by reactions with water. It is found in the gas and dust clouds floating in space and in planets, moons, comets, asteroids, and rocks which formed in the solar systems.

In the early days of the formation of our Solar System, the main repositories of water were the comets, asteroids and the rocks that formed the planets and other space objects. Comets are essentially packed snow, ice and dust that were part of the original gas cloud. They formed in the distant, very cold regions of the solar system. Asteroids are pieces of rocks, sometimes huge, which are mainly in orbit between Mars and Jupiter. We shall later find out how they came to be there.

There is still some discussion as to what was the main source of the water that we have on Earth. There are two main theories. One says that the water was delivered to the Earth by comets during the bombardment phase which ended about 3,800 million years ago. The bombardment by comets was very extensive in the early days of the Earth's formation. The distant orbits of some of these comets ('dirty snowballs') were disturbed by collisions with other space objects or by the gravitational attraction of Jupiter and other gas giants. Some orbits changed into ones that took the comets near to the Sun, crossing the orbits of the terrestrial planets. Some of these comets collided with the Earth and our Moon, some with the other terrestrial planets. Some were evaporated by the Sun while others continued their elongated orbits to the distant parts of the solar system to return many years later.

A lot of water was delivered to Earth by comets. In the early days of the Earth's life, before there was any atmospheric cover, much of the ice would have evaporated away. Analysis of cometary material brought back by space probes indicates that the Earth-water we have today is different at a detailed molecular level to the comet-water. So, the comets were probably not the major source of the water that we now have on Earth.

Another source of water was the asteroids that also rained down on Earth during the early bombardment period. There is a lot of water in space that exists within the rocks found there and combined in compounds within the rocks. The crashing asteroids would have delivered the water they carried.

Currently, the scientists favour the asteroids as being the main source of the water now on Earth.

A huge quantity of water was also present in the rocks that came together to form the planet. This water, together with the water that came with the comets and asteroids, meant that the Earth ended up with a large amount of water. The trouble was it was mainly trapped in rocks or underground.

But the early Earth was extremely hot, both on its surface and underneath. Over time, the heat inside the Earth changed the water trapped in the rocks to steam. The steam migrated to the surface through cracks in the crust. It evaporated into the atmosphere where it cooled, formed clouds, and condensed back into water. It began to rain. And it rained and rained. For years and years, perhaps for centuries, it rained onto the dry crust. The rain that fell onto the raised part of the crust soon found channels to flow to the lower levels. The channels deepened and became rivers. The rain that fell onto the depressions on the Earth's surface crust joined the water brought there by the rivers. The depressions filled up.

Remember that the part of the crust that rode above the depressions was the continental crust and the lower, denser part was the oceanic crust. Thus, when the rains finally stopped the higher continental crust was dry except for any flowing rivers and lakes, while the lower oceanic crust, now full of water, became the seas and oceans.

That is how the Earth ended up with dry land above sea level and oceans full of water.

Atmosphere

Apart from the water, many gases were also trapped in the rocks which came together to form the early Earth: nitrogen, carbon dioxide, methane, ammonia, and hydrogen sulphide (which smells of rotten eggs) were the chief components. There was little or no oxygen which is essential for life like ours to survive.

During the first billion or so years of the Earth's existence there was intense volcanic activity. Apart from the molten lava and solid rocks, the volcanoes sent up huge volumes of gases to the surface. The gases accumulated above the surface since the Earth's gravity stopped them from floating off into space. Slowly the first atmosphere formed.

You will note that the early atmosphere contained no oxygen and was poisonous to animal life; it was a while before life took hold. We will discuss in the next chapter how the atmosphere became suitable for life as we know it to begin and to evolve into the rich variety that exists today.

Before we continue our discussion of the other planets, let us get to know our Moon a little bit better.

Earth's moon: Moon

Fig. 5.17 Moon.

We have already seen above how scientists believe our Moon was formed, following the collision between the Earth and another planet, Theia, which was also in the process of forming. The Earth came out better in the collision, absorbing much of Theia. But it also lost material which then came together to form the Moon. The newly-formed Moon also suffered bombardment from asteroids and comets, which resulted in lava oozing through cracks in the crust on to the surface to form the lunar 'seas', seen as dark patches in Fig. 5.17, which are ancient lava flows, called *maria* (singular *mare*, pronounced /'ma:rer'/).

The end-result was that the Earth gained a new satellite. The new Moon was orbiting only about 25-30,000 km (15-20,000 miles) from the Earth when it formed but has been moving away gradually each year. Today, it is some 384,400 km (240,000 miles) from the Earth and getting further by about 3.8 cm (1.5 inches) per year, approximately the same rate that our fingernails grow. The Moon goes around the Earth once every 27.3 days. It rotates at the same rate as the Earth. This is because its orbit has become synchronised by Earth's gravity. The result is that we always see the same face of the moon. It was only when the US moon shots took astronauts around the moon that we got a glimpse of what lay on the opposite 'dark' side (dark for us because we had not seen it, rather than because there was no light).

The Moon is only the fifth largest of the moons in our Solar System. It is about a quarter of the size of Earth, with a radius of 1,737 km. Its gravity is about one-sixth that of the Earth. A 60 kg person will weigh only 10 kg on the Moon and will be able to jump six times as far and as high as on the Earth. The Moon has no atmosphere, and therefore, its days are extremely hot (127 C), and the nights exceedingly cold (-173 C).

[NB the ratio of the gravities of the Earth and the Moon depends on the ratio of the masses of the two objects].

Moon exploration

Fig. 5.18 Earthrise.

Chapter 5. Our neighbourhood…

The Moon is the only heavenly body that human beings have visited, apart from Earth of course. Neil Armstrong was the first human to walk on the Moon when he landed on the satellite on 20 July 1969 on the Apollo 11 mission together with Buzz Aldrin and said his famous words: "That's one small step for [a] man, one giant leap for mankind". Michael Collins stayed in the command module to await their return.

There have been six successful crewed missions to the Moon and a total of 12 astronauts have walked on its surface. The final Moon mission was Apollo 17 which landed on the moon on 19 December 1972.

A famous photo taken by the NASA Apollo 8 astronauts on Christmas Eve 1968 on the first trip round the Moon was named Earthrise [Fig. 5.18].

There has been a gap of almost 50 years since the last Moon landing, as I write. New manned missions are now being planned. The objectives are changing: a moon base may be established for a longer stay; the moon may serve as a staging post for a journey to Mars. Private enterprises are getting involved. Already SpaceX has taken astronauts to the International Space Station. Exciting times lie ahead.

Mars

Mars, the red planet, [Fig. 5.9, terrestrial planets], has fascinated mankind over the ages. It was named after the Roman god of war, no doubt because of its fiery red colour. It is the fourth planet from the Sun, and the first (and only) outer terrestrial planet for us – from Mars, the Earth would be seen to have phases, such as we see on Venus and the Moon. Mars itself has two moons Phobos and Deimos which are about 22 km and 12 km in diameter. Both are irregular in shape and probably are captured asteroids.

Mars has a radius of 3,390km which is just over half that of Earth (6,371 km). It is 228 million km from the Sun, which is about 1.5 AU, or 1.5 times as far as is the Earth. It orbits the Sun in 687 days which makes its year about 1.9 Earth years long. Its day is 24.6 hours long, much like ours. Its surface temperature can vary from -150 C to a balmy 20 C which is like a pleasant English summer's day.

Mars's internal structure is different to the Earth's. It has a dense core some 1,500-2,100 km in radius composed of iron, nickel, and sulphur. Above this is a mantle 1,240-1,880 km thick, and finally a crust 10-50 km deep made mainly of iron, but also containing magnesium, aluminium, calcium, and potassium. It is unlikely that Mars ever possessed a magnetic field.

We learnt a lot about Mars after the invention of telescopes. But what people first thought they saw through them has later not always shown to be there. Those who looked at Mars through the early, basic instruments saw that there was what appeared to be snow at Martian north and south poles. Then the Italian astronomer Giovanni Schiaparelli saw what he called *canali* (Italian for channels or canals) on the surface. Some people presumed that they were built by intelligent beings to bring water from the poles to irrigate the land. Legends began to grow that Martians lived on Mars. H G Wells wrote a book called "War of the Worlds" in 1897 about a Martian invasion of the Earth. [It is well worth a read if you have not done so already].

I am afraid none of these tales turned out to be correct, except, partially, the bit about snow on Mars. The Martian north and south poles do have snow, but the snow is not water-snow, rather it is frozen carbon dioxide [also called dry ice, it is used in theatres to create smoke since it evaporates readily and as it does, since carbon dioxide is heavier than air, it floats near the floor covering the actors' legs].

That apart, Mars is very dry with occasional dust storms, some of which can cover the entire planet. It gets its colour from its soil which has rust in it. Rust is what forms on the surface of iron when it has been in a damp location. What this means is that even though Mars does not have water now, it must have had it at one time.

We have already seen that flowing water is essential for life to exist. Thus, the possibility of water on the planet has led scientists to wonder if life exists or has ever existed on Mars. We shall discuss more about the possibility of life on Mars and our search for it in the next chapter. For now, let us say that water did once exist in abundance on Mars, and that it exists even now, but generally as ice under the surface. It is fascinating that there are visible signs on the surface of Mars that show ice sheets, and marks that indicate dried streams and other water flows [Fig. 5.19].

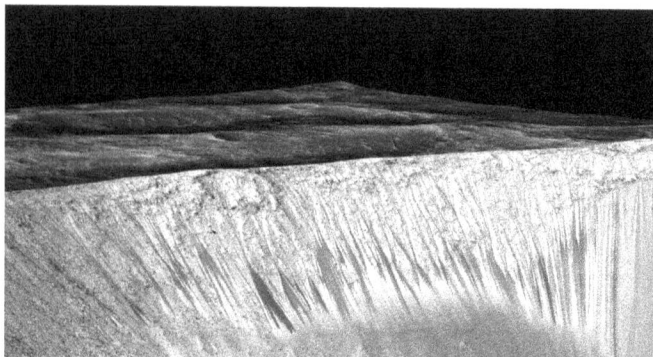

Fig. 5.19 Evidence of water flow on Mars.

Chapter 5. Our neighbourhood…

As a planet, Mars has several amazing features. On its surface it has Olympus Mons which is the highest and largest volcano in the Solar System, rising three times as high as Mount Everest in the Himalayas the highest peak on Earth. The volcano is named after Olympus, the mythical home of the Greek gods. Mars also has three other massive volcanoes near to Olympus Mons. The reason that all these volcanoes have been able to grow so big is that there is no plate tectonics on Mars: its crust is solid, mostly iron and not floating as is the Earth's. It is therefore able to carry the enormous weights of these massive mountains without sinking.

Another feature on Mars is Valles Marineris, a giant crack in the Martian crust. This is a canyon over 3000 km long, up to 200 km across and 7 km deep in places. In comparison, the Grand Canyon in the USA is about 445 km long, 30 km across and under 2 km deep.

In common with the other terrestrial planets, Mars has craters but fewer and more worn than the ones we see on Mercury and the Moon. The reason of course is that Mars has wind which can erode and fill in the craters with dust. Mars also has an atmosphere which produces the wind. The atmosphere is mainly made up of carbon dioxide, like that of Venus. However, it is very thin, being a thousand times less dense than that of Earth.

We have said above that Mars has signs that at one time, water flowed on its surface. For this to happen, the atmosphere must have been much thicker at that time: high atmospheric pressure is needed to stop the liquid water from evaporating away. What happened then to make Mars the dry, seemingly lifeless place that we find it today?

It is possible that the early Mars had a beginning rather like that of the Earth. The planet formed in a similar fashion and progressed to gases being expelled through the massive volcanoes that the planet possesses. The atmosphere that developed was mainly carbon dioxide, the greenhouse gas. For a period, the temperature on the planet was warm and water could exist as liquid. Rivers and large, shallow lakes, possibly even seas, could exist. But Mars is only about a quarter the volume of Earth, and its gravity is therefore much smaller; it would have been insufficient to stop the gas molecules drifting away from the planet into space. This loss of gas would have been made worse by the fact that Mars has had no, or perhaps only a weak, magnetic field as it has no churning core now and maybe never did. There was therefore no protection from the cosmic particles that would have bombarded the top of the atmosphere and knocked gas molecules off into space.

As the atmospheric cover was lost, ever more of the liquid water escaped, till today the only water left on Mars is underground in the form of solid ice. However, interestingly, occasionally the ice gets uncovered. If/when the temperature is warm enough (remember the temperature on Mars can get up to 20 C), the ice melts and runs down hills. How do we know this? Nasa's Mars Global Surveyor mission has taken photographs which show gullies left by water flowing down the side of a Martian crater [Fig. 5.19].

We know that liquid water is a key requirement for life. So, does this example of flowing water mean that life may exist today on Mars? We do not know, but it is unlikely. Still, could there have been life there at one time? Well, we do not know that for sure, yet.

Stop press: It was announced on 29 September 2020 that European Space Agency's (ESA) Mars Express spacecraft's MARSIS radar has discovered lakes of water under the south polar ice cap of Mars. There appears to be one main lake about 20 km (12 miles) across and at least 1 metre deep, plus at least three smaller surrounding bodies of water. It has been known for a while that water in form of ice exists under the Martian surface, but this is the first time that water has been identified in its liquid form. Scientists believe that liquid water can exist in the extreme cold because it is extremely salty. This would make normal life unlikely, even though bacteria have been found in similar lakes under the Antarctica ice sheet.

Asteroid belt

If we continued our journey away from the Sun, beyond Mars, till we are 300 million km (2.2 AU) from the Sun, we will come to the start of a very wide band of rocky, icy, and metallic bodies which stretches for some 150 million km (1 AU). This is the asteroid belt which spreads all around the Sun. The belt is composed of billions of objects, mainly irregular in shape, which range from tiny pebbles to bodies many kilometres in size. Some are big enough to have attained a spherical shape due to gravity. The chief amongst these is Ceres which orbits the Sun at about 414 million km (2.8 AU). Ceres has a diameter of 946 km and is spherical. It takes 4.6 Earth years to go around the Sun.

Ceres was discovered in 1801 by the Italian priest and astronomer Giuseppe Piazzi (1746-1826). It was granted a 'dwarf planet' status in 2006. It is the only object to be so nominated within Neptune's orbit. This also makes it the only object to be a dwarf planet and an asteroid at the same time. NASA's Dawn mission successfully reached Ceres in 2015.

Even though the asteroids in the belt are so numerous, their total mass is thought to be only about three times that of Ceres. It is believed that Phobos and Deimos, the Martian moons came from the asteroid belt and wandered near enough to the planet to be captured by Mars's gravity.

Where did the asteroid belt come from? One view is that the objects in the belt were on their way to forming a (small) planet when the massive gravity of Jupiter intervened and destroyed it. Today, the asteroids are of interest since they contain a lot of water and minerals of value on Earth, to the extent that plans are being considered with a view to capturing and mining them. On a more sombre note, asteroids pose a potential danger to Earth since they can crash on its surface with catastrophic effects to us humans. A close watch is kept on the biggest asteroids by astronomers. The big problem is the sheer number of these objects in space, since even some of the smaller ones have the potential to cause a lot of damage should they hit our planet.

Chapter 5. Our neighbourhood…

The gas giants

Beyond Mars and the Asteroid Belt there are four planets we call the gas giants [Fig. 5.20]. First, we find the mighty Jupiter, the king of our planetary system. Beyond Jupiter is the beautiful Saturn, then the distant Uranus and finally Neptune.

How did these planets form?

Once the Sun formed, its solar wind was blowing the gases in the cloud away from the centre of the Solar System. Perhaps in only about 5 million years after the Sun was born, all the gas and dust in the disk had been blown away. The gas giants would have had to form extremely quickly. It is still uncertain as to how this happened.

The further you go from the Sun the weaker the solar wind becomes and the colder it gets. So as the Solar System was forming, the distant parts of the cloud were cold enough to retain the gases, and for ice to exist.

The gas giants could have formed near the Sun where there was plenty of gas as well as rocks to build the core. But how did they move to the distant parts of the Solar System where they are now found? Or could these giants have formed rapidly far from the centre, while there was still plenty of gas and ice around? In the latter case, they are likely to be all gaseous without any rocky core. The jury is still out on this point; the answer depends on finding out whether the gas giants have a rocky core or not.

Fig. 5.20 The four gas giants.
The planets from right to left:
Jupiter, Saturn, Uranus, and Neptune.

If the core is not rocky, it is probably made of hydrogen which is solid at the extreme cold of the far space and the enormous pressures at the centres of these huge objects. The gas giants may well have a core of hydrogen ice as well as of rock. In any case, with plenty of gas (mainly hydrogen and helium, remember) available, they were able to grow extremely large.

Jupiter

Jupiter [Fig. 5.20] is the big daddy of the planets in the Solar System. It is two and a half times as heavy as all the other planets in the Solar System combined. It is big enough in size to fit a thousand Earths within it. In fact, the smallest star we know of, a red dwarf called OGLE-TR-122b, is only 20% bigger than Jupiter; it is however more than 100 times heavier than the planet. Mass is the important thing that determines whether a star can start the nuclear reactions which are necessary for it to shine. Thus, there was never a possibility of Jupiter becoming a star. However, interestingly, Jupiter does produce more energy than it receives from the Sun and radiates this out into space.

There are many other intriguing points about this giant. Jupiter rotates faster than any other of the Solar System planets, spinning about its axis in less than 10 hours. Jupiter's day is therefore 14 hours shorter than our own. But since it is so far from the Sun – more than 5 times as far as is the Earth – its year, the time it takes to orbit the Sun, is almost 12 Earth years. Because it is rotating so fast, and because of its size and the currents within its gaseous interior, Jupiter has huge cloud belts and storms on its surface. In fact, one storm, called the Great Red Spot due to its colour and size (you could fit three Earths within it), has been raging on Jupiter, to our knowledge, for more than 350 years! It was probably first seen in 1665 by Giovanni Cassini (1625-1712), the Italian/later-French astronomer and mathematician who also discovered four of the satellites of Saturn.

Jupiter is the fourth brightest object in the sky after the Sun, Moon and Venus, and is clearly visible to the naked eye. Several space missions have been to the planet and its moons. These include Pioneer 10, Voyagers 1 and 2, Galileo, Cassini, and New Horizons. There are likely to be many more in the future.

The moons of Jupiter

Jupiter has 67 moons at the last count: more than any other Solar System planet. Four of these moons are big enough to be easily seen through even a small telescope.

The moons were first noted by Galileo Galilei (1564-1642), an Italian astronomer, mathematician, and man of science. Let us take a short deviation to talk about this remarkable man.

Chapter 5. Our neighbourhood...

Galileo who was born in Pisa in 1564 was a genius. He was involved in many developments in science including the laws of physics relating to the motion of objects. He became fascinated by telescopes that had recently been invented. He soon learnt all about them, made them himself and improved their design. He studied the Moon and the planets, soon discovering that Jupiter had four smaller objects that went around it. These were of course the four major moons of Jupiter. This discovery got him into trouble with the then powerful church authorities.

At that time, it was believed that the Earth was the centre of the Universe and all objects in the sky revolved around it. You can understand why this was so because if you looked up in space everything, including the Sun, appears to go around the Earth. The Church believed that the stars and planets were spinning on different shells ('heavenly spheres') around the Earth in the centre. This was first imagined by Ptolemy (100-170) and was still believed by the church in the 16th century some 1400 years later.

This was the church's vision of perfection. So, when Galileo started talking about the moons of Jupiter going around the planet, this did not fit into their scheme: if the moons were going around Jupiter, they could not be going around the Earth.

Galileo was arrested, tried, and found guilty. He was given life imprisonment. Eventually, he was forced to confess he was wrong and that the Earth did not move but everything else did, around it. He was released, but it is said that he muttered under his breath, referring to the Earth, "And yet it moves". He knew he was right as history has proved.

Galileo is also famous for his demonstration at the leaning tower of Pisa where he dropped heavy balls and showed that objects fall at the same rate [strictly, in a vacuum] irrespective of their weight. He died in 1642, leaving us a wealth of knowledge because of his work.

Anyway, let us get back to Jupiter's moons.

Fig. 5.21 The four main Jovian moons.
The four shown from the left, are: Io, Europa, Ganymede and Callisto.

The four main moons of Jupiter are Io, Europa, Ganymede and Callisto [Fig. 5.21]. Each of these moons is fascinating. Let us take a brief look at them, starting with Io, the one nearest to Jupiter.

Jupiter's moon: Io

Io [Fig. 5.21] is a little bit bigger than our own Moon and is the second smallest of Jupiter's four moons that we are discussing. It looks rather like a pizza with its blotchy, colourful surface. The reason for this look is that Io has more than 400 volcanoes on its surface. In fact, it is the most active object in the Solar System. The gases and material that are spewed out are very sulphurous; they would stink of rotten eggs and be poisonous to us. They coat Io's surface in yellows, greens, and browns.

It is interesting to consider why Io has any volcanoes at all. It is so small that any heat in its interior should have been lost long ago, as happened to our Moon. For there to be volcanoes, there needs to be molten material underground which can be spewed out of surface cones. However, for the material to be molten, there needs to be heat, enough to melt the material. Way out in space where Jupiter and the other gas giants are found, the temperature is so low that apart from some of the gases, everything is frozen solid, especially in a small object like Io. So where is the heat coming from?

Now I want you to try an experiment. Take a soft rubber ball. When you touch it, you will find that is not hot, or even warm. Now squeeze the ball in your hand then release the pressure and allow the ball to return to normal. Do this action many times. After a while you will find that the ball has become warm. This is due to all the work that you have done in pressing and releasing the ball. Much of the energy you spent has gone into raising the ball's temperature.

The orbit of Io, as it goes around Jupiter, is affected not only by Jupiter but also the other moons. The result is that Io's orbit is not circular. The moon comes extremely near to Jupiter and then moves far away. When it is near Jupiter, it is squeezed by the huge gravity of the giant planet. This squeezing becomes less when the moon is further away from Jupiter and again increases when it is near. Io is like a rubber ball in Jupiter's hand. Its interior gets heated, and melts. This is called tidal heating. The molten stuff in Io's interior gushes out to heights up to 200 km through the hundreds of volcanoes on the moon and coats the surface with sulphur and other substances.

[The gravitational pull between a planet and its moon is the same effect that causes the Moon to bulge slightly on the side facing the Earth and gives us the tides in our seas].

Jupiter's moon: Europa

Europa [Fig. 5.21] is the smallest of the four moons of Jupiter we are discussing, being about the same size as our Moon. Its orbit is very nearly circular, but not totally so. You would therefore expect that it too gets affected by Jupiter's tidal heating; you would not be wrong. In Europa's case, we think that there is a small rocky core and enough tidal heating to melt the ice in the interior, resulting in a huge ocean under the surface ice cover. This water escapes through cracks in the icy surface and freezes on top. As you can see from the picture of the moon, the result is remarkably like the surface of the Arctic Ocean over the Earth's North Pole in the winter with its floating ice sheets.

When we talk about how life began on Earth, one of the likely ways is thought to be that it started under the sea, where there are hot vents circulating material from deep under the sea floor [see Fig. 6.3]. There is a possibility that the same thing could be happening at the bottom of the hidden ocean under Europa's ice cover.

Europa is one of the places with the likelihood of us finding extraterrestrial life in our Solar System. Expeditions are under serious planning to visit this moon of Jupiter in the not-too-distant future. NASA has been working on conceptual proposals to explore this region; a report on one option was produced as long ago as 2013. I am sure we will hear a lot more about this fascinating satellite in the future.

Jupiter's moon: Ganymede

Ganymede [Fig. 5.21] is the largest of Jupiter's moons. It is much larger than our Moon, and indeed it is larger than any moon in the Solar System; it is even larger than the planet Mercury. It has a magnetic field like the Earth and is believed to have a molten iron core like the Earth. Like Europa, Ganymede's surface is icy and probably has a water ocean beneath it. Its surface has light and dark areas with the latter seeming to be covered with organic material.

Ganymede was probably formed near Jupiter at the same time as the gas giant. It has a thin atmosphere. The Hubble Space Telescope found that this contained oxygen which is believed to have been created by the water in the surface ice being split into its components, hydrogen, and oxygen, perhaps by ultra-violet (UV) radiation from the Sun.

Jupiter's moon: Callisto

Callisto [Fig. 5.21] is the last of the moons of Jupiter we are discussing. It is the second largest, after Ganymede, and is the furthest of the four from Jupiter. Its surface is covered with water ice, frozen carbon dioxide (dry ice), particles of rocks and hydrocarbons (these, as we know, are molecules made of hydrogen and carbon which are the basis of organic, living matter.)

Callisto has no detected volcanic activity. However, its surface is the most heavily cratered of any object in our Solar System. We think that deep under the hard surface there is an ocean of salty water ice as on the other Jovian moons. There is a thin atmosphere on the moon which chiefly contains carbon dioxide.

Saturn

The next planet after Jupiter going away from the Sun is Saturn [Fig. 5.20], perhaps one of the most beautiful objects in our night sky. It is well known for its spectacular and prominent rings. If you get a chance to look through a telescope at Saturn, take it. At least, look at its images on the internet. The image of the planet with open rings is one that has inspired many a young person to take up astronomy.

Saturn is as big as 95 Earths. It orbits the Sun at 9.54 AU (that is, it is 8.54 AU further from the Sun than the Earth) and takes 29.5 Earth years to do it. It spins on its axis in just 10 hours 34 minutes, only a little slower than Jupiter. It is composed almost entirely of hydrogen. This makes it extremely light. Its density is so low in fact that if you could put it in an ocean of water, it would float. Being so light and spinning so fast means that it is squashed rather like an orange, and bulges at the Equator – its diameter at its equator is 120,536 km while its polar diameter is only 108,728 km. Saturn's upper atmosphere has banded clouds, with many oval-shaped storms rather like those on Jupiter's surface.

Saturn has over 62 moons (more keep getting discovered), but over 150 if the small moonlets are included.

Saturn has been extensively studied by the various space missions that have visited it, including: Pioneer 11, the two Voyagers and Cassini-Huygens. You can find the many photographs and other achievements of these missions quite easily on the internet.

Fig. 5.22 The day the Earth smiled.
The Earth from Saturn:
the pale blue dot.

Chapter 5. Our neighbourhood…

One of the most poignant images we have of our planet was taken by the Cassini mission when it turned its cameras towards Saturn and caught the Earth as a tiny blue dot 900 million miles away [Fig. 5.22].

The rings of Saturn

Saturn has more than 30 rings with gaps between the rings, arranged in 7 Groups. Two of the groups have moons within them shepherding or guiding the rings and keeping them in place by their gravity. Other gaps have small moonlets within them. The rings are made up of billions of rock and ice pieces ranging from dust- to boulder-sized. The rings stretch out over 120,000 km from the surface of the planet. The amazing thing is that the rings are in general only about 10 to 20 meters thick, though bulging in parts.

The origin of the rings is uncertain, though they are likely to be moons or captured asteroids and comets, broken up by Saturn's gravity.

The moons of Saturn

We have said that Saturn has at least 62 moons. Here we will consider three of the more interesting ones: Titan, Enceladus, and Iapetus.

(a) Titan.

(b) Enceladus showing the geysers.

(c) Iapetus showing its two sides.

Fig. 5.23 Three of the moons of Saturn.

Saturn's moon: Titan

Titan [Fig. 5.23 (a)] is Saturn's largest moon and the second largest moon in our Solar System; only Ganymede is bigger. Its diameter is 5150 km compared to our Moon's 3474 km. It goes around Saturn in about 16 days. It is a remarkable place; in many ways it is like the Earth was soon after it formed.

Its atmosphere is thick and foggy, orange in colour and made of methane, ethane, and other organic compounds. The 'air' is heavy enough to stop us seeing any of its surface. However, on January 14, 2005, the Cassini-Huygens mission successfully dropped the Huygens probe onto Titan's surface, which sent back data including images of the surface, including images of the surface seen as it was descending, and of its landing area.

The surface turns out to have hills, valleys, riverbeds, and lakes. It seems familiar to us on Earth, except that Titian's temperature is -179 C, and when it rains liquid methane and ethane fall from the thick clouds and flow down to the lakes which are filled with these liquid hydrocarbons. There are volcanoes on Titan, but these are cryovolcanoes which spout out super-cold water and other ices.

It is believed that Titan has a rocky core which may still be hot, surrounded by liquid water and other substances such as ammonia. It probably formed at the same time as Saturn and was near enough to be captured as a satellite of the gas giant.

Titan has the potential for harbouring life, if not now, certainly in the far future when the Sun bloats and gets near enough to warm the atmosphere.

Titan was discovered in 1655 by Christiaan Huygens (1629-1695), the famous Dutch mathematician, astronomer, and physicist. Huygens also found out the true shape of the rings of Saturn. The probe that landed on Saturn's moon Titan was named after him.

Saturn's moon: Enceladus

Enceladus [Fig. 5.23 (b)] is the sixth largest moon of Saturn but is only 504 km in diameter and orbits the planet in only 1.4 days. It was discovered in 1789 by William Herschel (1738-1822), a renowned British musician and astronomer of German origin. Herschel also discovered the planet Uranus in 1781. His sister Caroline was also a successful astronomer being the first woman to discover a comet.

Enceladus may be tiny, but it is important and has been extensively studied. It is an icy world with a rocky core and water ice under the frozen crust. The interesting thing about Enceladus is that it has over 100 geysers in the 'tiger stripes' on its surface near to its south pole, as can be seen in Fig. 5.23 (b). These occur due to tidal heating caused by its orbit around Saturn, or possibly due to radioactive heating in its interior, and send spouts of water ice particles out into space. Some of these particles land back onto Enceladus but some are carried across to feed the E-ring on Saturn.

The Cassini spacecraft took an opportunity and flew low through these geysers and collected the liquid for analysis. When the instruments on board examined the material, they identified liquid water and organic compounds. This has excited the scientists, and now Enceladus is amongst the top contenders (together with Jupiter's moon Europa) to harbour extraterrestrial life in our Solar System. It is believed that Enceladus has a liquid ocean under its surface.

Saturn's moon: Iapetus

Iapetus [Fig. 5.23 (c)] is interesting because of its odd shape and surface features. It is 1471 km in diameter (compared to our Moon's diameter of 3474 km), and orbits a long way from the planet taking over 79 days to travel the 3,560,850 km. It was discovered in 1671 by Giovanni Cassini of whom we have spoken earlier.

Iapetus is shaped rather like a walnut and has a strange mountain ridge 13 km high (Mount Everest is only 8.8 km high) running 1300 km across the moon. Even stranger, one side of Iapetus is bright while the other is dark (as seen in two pictures of the moon in Fig. 5.23 (c)). Astronomers do not fully understand the reasons for this. However, they believe that the ices in the warmer part of the region evaporate leaving behind the darker organic material.

Iapetus is heavily cratered with a huge impact area of 580 km diameter on its surface, as can be seen in Fig. 5.23 (c).

Uranus

Uranus [Fig. 5.20] is the seventh planet from the Sun. It is 15 times as massive as the Earth, and about 50,000 km in diameter. It is 19.8 AU (almost 2.9 billion km) away from the Sun and takes 84 Earth-years (its year) to complete a circuit. It is a lovely pale blue in colour.

There are a couple of strange facts about Uranus apart from its distance from the Sun and cold surface (-216 C). Uranus rotates in the opposite (retrograde) direction to most of the other planets, taking 17 hours and 14 minutes per rotation. The only other planet that does this in our Solar System is Venus. But even stranger is the fact that Uranus is tipped over on its side and so rolls along its orbit, perhaps because of a collision when it was forming which could also have been responsible for the retrograde rotation. This means that its poles often point directly at the Sun with the result that each pole is in light for 42 years (half of its year), and in darkness for 42 years. In comparison, the Earth's poles are in light or darkness for 6 months.

The surface of Uranus can get as cold as – 224 C which makes it one of the two (the other is Neptune) coldest planets in the Solar System. Its top surface layers are made mainly of hydrogen, with a little helium. They lie above an icy layer below which is a core of rock and ice. The planet is surrounded in an atmosphere of water ice, ammonia, and methane ice crystals, which hide any cloud features or storms on the surface. This atmosphere is itself surrounded by a methane haze which gives Uranus its colour.

Another interesting fact about Uranus is that it has faint rings (not seen in the Fig. 5.20 image), 13 in total, orbiting the planet in two groups: the inner group has 11 rings, and the outer group has two. The probable cause of these rings is believed to be the shattered remains of a moon which broke up.

Uranus has 27 moons, frozen and dark. All are named after characters from Shakespeare's plays. The most interesting one is Miranda which has got cracks and 20 km high cliffs on its surface. The biggest is Titania which is the eighth largest in the Solar System.

As we noted above, Uranus was discovered by Sir William Herschel in 1781, who also went on to discover Enceladus eight years later.

Neptune

The final planet in the Sun's Solar System family is Neptune [Fig. 5.20], a cold distant object with a spectacular bright blue colour.

It is interesting to compare the facts about Neptune and Uranus.

Neptune is more than 1.5 times as far from the Sun as Uranus. Each orbit of Neptune's takes a 164.8 Earth-years to complete compared with Uranus' orbit of 84 years. Neptune orbits the Sun at almost a staggering 4.5 billion km (30.1 AU) distance, which is more than 30 times as far as the Sun-Earth distance, compared with the radius of Uranus' orbit of about 2.9 billion km (19.19 AU). Neptune takes about 18 hours to rotate about its axis, again a similar rate to Uranus' 17 hours 14 minutes. However, while Uranus has a retrograde rotation, Neptune's rotation is normal. Neptune is almost the same size as Uranus, about 49,000 km in diameter, and just a little heavier (17 times as massive as the Earth against 15 times in the case of Uranus).

Neptune has a very thick atmosphere. It is made of hydrogen and helium, with some methane. Methane absorbs red light and that is what gives Neptune its blue colour. Photos show wispy white clouds floating in the upper atmosphere. The planet has very high-speed winds whipping around the planet at more than 3000 km per hour, or almost 2000 mph. Compare this with the 150 miles per hour winds we experience here on Earth in a bad hurricane. A huge storm, called the Great Dark Spot which was seen by the Voyager 2 space probe on the planet in 1989 lasted a good five years. (But remember the Great Red Spot storm on Jupiter which is still going strong after more than 350 years). Below the upper atmosphere, there are layers of hydrogen, helium, and methane gases, followed by water, ammonia, and methane ices. At its centre, the planet has a rocky core.

Like Uranus, Neptune has a ring system (again too faint to be seen in the Fig. 5.23 image). In Neptune's case, there are 5 very thin rings made up primarily of ice and dust particles.

It is interesting how Neptune was discovered. Astronomers found that the precise orbit of Uranus was slightly different to one that was predicted by scientific analysis. They concluded that its orbit was being affected (or 'perturbed', in exact terminology) by another body orbiting further out in space. They calculated where the body would be found and looked and discovered Neptune exactly where they had predicted it to be!

There are 13 moons orbiting Neptune with the main one being Triton, which we will now review briefly.

Neptune's moon: Triton

Triton is an icy object about a third of the size of our Moon which is orbiting Neptune in a retrograde direction, opposite to that of the planet's rotation, once every 5.9 days. This probably means that Triton is an object that came too near to Neptune and was captured by the planet's gravity.

Triton's surface is very varied with uneven, icy surface features. There is also evidence of geysers, which spout nitrogen gas 8 km above the surface. Because of the geysers, Triton has a very thin nitrogen atmosphere. The surface also indicates that other material has erupted from the moon's interior. This in turn shows that there is cryovolcanic activity and thus heat in the interior of the moon, probably due to tidal heating caused by the pull of Neptune's gravity. [Note: Cryovolcanoes eject water and other liquids such as methane at extremely low temperatures rather than hot molten rock].

The outlook for Triton is not too bright. It appears likely that the moon will move progressively closer and closer to Neptune till the planet's gravity breaks it up in about 3.5 billion years. Neptune will then gain a new ring system.

Voyager 2 has been the only spacecraft yet to have flown by Neptune and Triton (in 1989).

The visit to Neptune brings us to the end of the Sun's planetary family but not to the end of the Solar System itself. Two further features lie beyond that we need to talk about: the Kuiper Belt and the Oort Cloud.

The Kuiper Belt

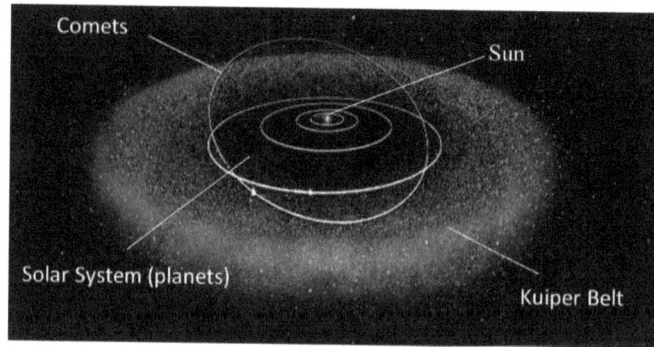

Fig. 5.24 Kuiper Belt.

The Kuiper Belt [Fig. 5.24] is a region of space stretching from about 30 AU, where Neptune orbits, to 50-70 AU from the Sun. This region is home to icy worlds and objects ranging from small pieces of ice to Dwarf Planets with diameters up to the 1000+ km. The Kuiper Belt is also the origin of comets with short-period (up to about 200 year) orbits round the Sun. The ices found in this far region of space consist of water ice and frozen volatile gases (volatile gases are those that would be vapours at the normal temperatures found on Earth; these include methane, nitrogen, and ammonia). These ices are debris left behind from the formation of the Solar System, and therefore date back some 4.5 billion years. Objects in this space are called Kuiper Belt objects (KBOs), or sometimes Trans-Neptunian Objects.

The Kuiper Belt (also called the Kuiper-Edgeworth Belt) was named after the Dutch American astronomer Gerard Kuiper (1905-1973) who developed a theory that a disk of material existed in the far reaches of space, and the Irish astronomer, economist, and engineer Kenneth Edgeworth (1880-1972) who predicted that there are likely to be icy objects left over from building the Sun and the planets.

The Dwarf Planets were a category created in 2006 when several new 'Pluto-sized' objects were found. It was decided to restrict the title of a "planet" to the eight major ones ranging from Mercury to Neptune and to label all smaller contenders as Dwarf Planets. There are four main Dwarf Planets to be found in the Kuiper belt: Pluto, which was once a main planet but was downgraded in 2006 to the status of a dwarf planet, Humea, Makemake and Eris. The fifth dwarf planet Ceres lies, in fact, in the Asteroid Belt. Ceres, therefore, orbits the Sun in only about 4.6 years, compared with the lengthy orbit times of the others: Pluto in 248 years, Humea in a little over 283 years, Makemake in almost 310 years and Eris in about 561 years.

Chapter 5. Our neighbourhood...

Dwarf planet: Pluto

Fig. 5.25 Pluto.

Pluto [Fig. 5.25] is the largest of the Dwarf Planets and yet is smaller than our Moon. It has a thin atmosphere of nitrogen, and a methane haze surrounding it.

Interestingly, the orbit of Pluto is so elliptical that it takes it close enough to the Sun to thaw the surface ices and allow an atmosphere to form. The stronger sunlight at the near points to the Sun also breaks down the atmospheric methane into the hydrocarbons of which it is composed.

When Pluto is far from the Sun, the intense cold of deep space refreezes the methane and hydrocarbons in the atmosphere, which fall back onto the surface and give it a dark coating.

The New Horizons mission visited the Kuiper Belt in 2015 and flew by Pluto and its moon Charon that year. These two turned out to be more exciting than expected.

Pluto was found to have a range of mountains made of water ice. Charon has developed a red coloured 'north' pole through stealing Pluto's methane atmosphere over the ages and depositing it on its pole at an exceedingly slow rate of about 1mm every million years. Since Charon is inactive, the frozen gases have stayed where they fell.

The Oort Cloud

Some 2000 AU (about 300 billion km or 300,000,000,000 km) away from the Sun and stretching for some 100,000 AU (15 trillion km or 15,000,000,000,000 km, about 1.5 light-years!) is believed to lie a huge spherical cloud of icy objects called the Oort Cloud [Fig. 5.26] that envelopes the Solar System including the Kuiper belt.

The existence of this cloud was developed in a theory by Jan Oort (1900-1992) a Dutch astronomer, and the cloud was named after him. Oort also showed that our Milky Way galaxy rotates and introduced the idea of dark matter.

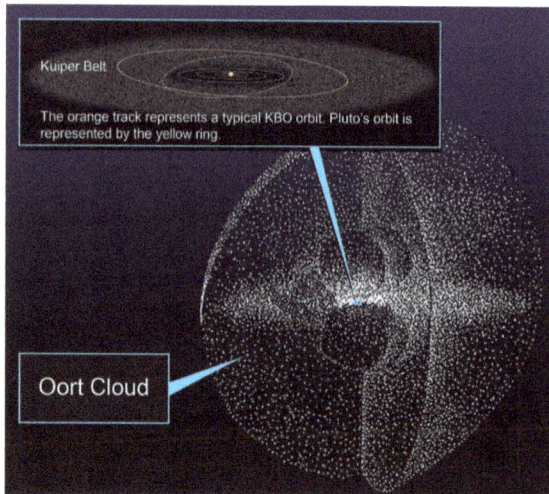

Fig. 5.26 The Oort Cloud.
KBO refers to Kuiper Belt Objects.

The cloud is thought to be the remnant of the objects that formed the Solar System. After the formation of the planets, the 'unused' icy particles near to the Sun were flung by the giant Jupiter's gravity to the outer reaches of the Solar System. The Oort Cloud is far enough from the Sun to be influenced by any nearby passing star or nebula which can cause some of its objects to be knocked out of stable orbits and sent hurtling towards the Sun as comets.

[An alternate recent theory says that there was a twin to our Sun, and that these two stars gathered space debris up into what became the Oort cloud. Later the twin sun wandered away from ours. An interesting fact is that while virtually everything else in our Solar system orbits the Sun in the plane of the original dust cloud that gave birth to the Solar System, the Oort cloud alone is spherical, surrounding the rest of the Solar System. The twin suns theory would address this anomaly.]

The Oort Cloud is believed to be the source of comets with over 200-year orbits. One long-orbit comet, which made a close pass of Mars in 2014, is not expected to return for 740,000 years.

Have we now reached the end of the Solar System?

If by the 'end of the Solar System' we mean the region where the influence of the Sun's solar wind collides with the 'interstellar medium' (material in the region between the stars), the answer is yes. This region is called the 'heliopause' [Fig. 6.6]. We are not certain, but it is believed to start about 120 AU (approximately 18 billion km) from the Sun. Voyager 1 passed through this region at the end of 2004 and Voyager 2 in 2007.

However, if by the 'end of the Solar System' we mean the region where the influence of the Sun's gravity ends, then we need to go beyond the Oort Cloud to about 2 light-years from the Sun, almost halfway to Proxima Centauri the star nearest to the Sun. Objects within this distance will orbit around the Sun.

What next for the Solar System?

We have seen that the Solar System was born following the collapse of a huge, cold cloud of gas and dust which was drifting in space till it was given a shove by the explosion of a nearby supernova. We noted that this resulted in the creation of the Sun at the centre of the collapsing cloud, and of the planets – both major and dwarf – by the coming together of the dust particles in the cloud circulating around the Sun. We observed that apart from the planets, the Solar System is home to asteroids, comets, and icy and rocky bodies, some minuscule, others large, that were left behind after the Solar System's planet building was completed. Finally, we saw how big the Solar System is, with the influence of the Sun stretching half-way to its nearest neighbouring star.

But what is the fate of the Solar System? What will happen to the Sun and the planets and all the other objects spinning around it? What will become of the Earth?

One thing we know for certain is that nothing lasts forever. The same is true of the Solar System. The clue to knowing its fate lies in knowing what happens to the Sun.

We read in chapter 3 on Stars that all stars die, and how. We saw that the Sun is a star, and it too will die one day. We know approximately when that will occur and how it will happen. We know that how a star dies depends on how big it is. The Sun is an average mass star. Such stars shine by 'burning' the hydrogen in their core to produce helium. When the hydrogen fuel is finished, the core collapses and helium burning starts converting the helium to form carbon. While the helium is burning, the star expands till it is 10 to 100 times its original diameter. While the star's surface expands, its core collapses till it is only a few times the size of the Earth.

As the star expands, its surface cools down as its energy is spread over a larger surface, and its colour changes from yellow to red. It becomes a red giant. When the star is a red giant, its gravity is unable to stop the material in its outer layers drifting away. Slowly the star 'evaporates' till all that is left is a glowing white core called a white dwarf. Eventually, the core cools over many billions of years till all that remains is a burnt-out cinder, a black dwarf. That is the fate of the Sun.

The Sun was born about 4.5 billion years ago and it is about half-way through its life. It will show signs of its approaching end in about 3.5 to 4 billion years. But what will be the effect of the Sun's fate on the rest of the Solar System?

The Sun will increase over 100 times in size to become a red giant, shedding perhaps half its mass in a nebula. It will continue to expand and may have a radius of 1 AU (the distance to the Earth) or even more. As it balloons, it will swallow up Mercury and Venus, and possible even the Earth. If it survives, Earth would have been burnt into a cinder and its seas evaporated. Any life existing would die. But it is unlikely that there would be any life left on Earth by this time.

There is another, earlier danger for life on Earth. The Sun is progressively getting hotter and brighter with time. To begin with, the Sun was only about 70% as hot as it is today. Over the last 4.5 billion years, it has reached today's temperatures. As time passes over the millennia, the increasing temperature of the Sun will push the habitable Goldilocks zone further away. Remember, this is the area where liquid water can exist on the surface of a planet. Once the zone goes beyond Earth's orbit, our seas would have boiled or evaporated away. No life can survive this. It is anticipated that this situation could arise 1.75 to 3.75 billion years from now. By then we would hope that mankind, if it has survived in some form or the other, would have long left planet Earth for another world which is still within its star's Goldilocks zone.

Of course, as the habitable zone moves further out into space, it will reach Mars the outer terrestrial planet in the Solar System. This could be the next destination for our civilisation. Perhaps in due course, Titan will come into its own and convert into a life-supporting object. But in time, even Mars and Titan will not be habitable, and the Solar System will lose all capability to support life. Perhaps, from our life-centric point of view, this is the time when we can say that the Solar System has well and truly died.

The Sun itself will eventually finish all its fuel, shed almost half of its mass in a nebula and turn into a white dwarf. The smaller, lighter Sun's gravitational hold on the planets would loosen and their orbits become erratic. Some may drift away into the cold space, while some may collide, or fall into the tiny Sun, and destroy themselves.

For countless trillions of years beyond, the planets that remain will continue to orbit the point in sky where the white dwarf sits losing energy, slowly turning into a black dwarf.

All this will happen in a frozen, lifeless darkness.

Chapter 6

Are we alone...?

where is ET when you need him?

Earth is home to us, and to millions of other living beings: big and small, ones that live on land and underground, in the sea and in the air, and inside other creatures. Some living things on our planet are so tiny that they can only be seen with powerful microscopes while others are the largest creatures that ever lived.

Earth is simply teeming with life. But Earth is also the only place in the Universe that we know where life exists. So far.

As we look around our home, and up at other stars in the sky, there are lots of questions that come to mind.

How did life start on Earth? How did it evolve? Will it last forever? How will it end? Are we alone? Or are there other worlds where life may exist? Where are they? Are they near, or in far-off solar systems? Is there life in other worlds in our Solar System? How will we find it? Will alien life be intelligent like us or simple like bacteria? What are the chances of there being other intelligent civilisations in our galaxy? In the Universe? Will the extraterrestrial civilisations be technologically advanced? If they are more advanced than us, what will that mean for us? Should we be actively seeking these civilisations out, or simply be listening for their communications?

Questions. Questions. This is what this chapter is about: to try as best as I can to give you some answers.

Life on Earth

The Earth was formed some 4.6 billion years ago. Scientists think that within eight million years after this, and possibly earlier, life had established itself here. The question is how did it start? This is uncertain but, as you should have learnt to expect by now, there are theories.

One thought is that life, or at least many of the complex molecules (called organic molecules) needed for life as we know it, were assembled in outer space, and brought to Earth within comets and/or in asteroids. This is not as bizarre as it sounds.

Some of the complex molecules that are found in living cells, have been discovered in comets and in asteroids. These molecules would have landed on the primitive Earth, where the conditions were quite different to those we find today. It would have been much hotter at that time, and pools of water rich in chemicals, and storms, and lightning strikes would have been quite common. It was in such a surrounding, that the first living cell could have formed.

Experiments on Earth where the essential chemicals needed for these molecules have been put together in a beaker, and an electric current passed through them to simulate a lightning strike, produced a gooey, dark, sticky substance which contained some of the organic molecules that were discovered in the material from space. They could have been produced naturally on Earth.

We will first look at DNA molecules which are needed for life, then how see how life could have started and evolved.

The structure of life – the DNA molecule

[You may skip this section and return to it later]

All animals and plants on Earth are made up of *eukaryotic* cells. Such a cell consists of a nucleus containing genetic material (information about controlling the functioning of the cell) and a jelly-like substance (called cytoplasm). The nucleus and the cytoplasm are enclosed by a covering called a membrane. These cells are tiny; the human body has an estimated 30-40 trillion of them. The human cells are extremely specialised, there are cells that make up your eyes, your teeth, build and repair organs, there are stem cells that can change into any other type of cell, etc. What they are and what they do is controlled by the information within their nuclei. This information is carried by especially important molecules called DNA.

DNA is shaped like a ladder which has been twisted around as if in a spiral staircase (the shape is called a double helix) [Fig. 6.1 (a)]. The steps of the ladder are made of two molecules which together are called 'base pairs'. Each of the base pair molecule is called a nucleotide. The nucleotides themselves come in one of four forms ('A', 'C', 'G' and 'T'). The 'A' nucleotide always pairs with 'T' and 'C' with 'G'.

Chapter 6. Are we alone…?

Human DNA has some 3 billion base-pairs. Genes are groups of base pairs which form a DNA molecule. Genes determine everything about us that make each of us unique: our height, the colour of our eyes, etc.

Proteins are complex molecules which are the building blocks of your body making your muscle, bone, and other organs, and carrying out essential tasks such as repair and regulating the function of the body's tissues and organs.

A DNA molecule together with a protein is stored, tightly packed, in our cells in strands called chromosomes. Each person has 22 pairs of these chromosomes plus 2 sex chromosomes (XX in females and XY in males) making a total of 46. The base-pairs in our DNA are instructions for the cell to produce specific proteins to carry out specific functions.

The finding of the DNA structure was a major milestone in science. Its discovery is a story worth telling (and reading) in its own right. Here, let us just name some names.

The 1962 Nobel Prize in Physiology or Medicine for the discovery of the DNA structure was awarded to three scientists: Francis Crick (1916-2004), a brilliant British biologist, James Watson (1928-), an impetuous American zoologist and geneticist, both working at the Cavendish Laboratories at the University of Cambridge, and Maurice Wilkins (1916-2004), a New Zealand born British physicist and microbiologist, for their discovery of the structure of DNA.

Another major contributor to the discovery, Rosalind Franklin (1920-1958) was not so honoured. The Nobel committee's decision was controversial. I cover it in more detail in Appendix C under the story of Rosalind Franklin.

[End of optional section]

How cells divide and evolve

We have seen above that all living cells contain instructions for making the cells and for managing their functions. These instructions are coded into incredible molecules called DNA.

We start life in our mother's womb as just one cell, but are born as babies, fully formed little people; we grow to become children and finally develop even more to become adults. Our cells must multiply to allow all this growth and change to happen. How do they do this?

Living cells have an amazing capability – they can divide into two identical cells. One cell can become two, two can become four, four become eight, on and on. Thus, for life to begin, we needed the first cell to be created, not just any cell, but one that could split itself into two identical cells.

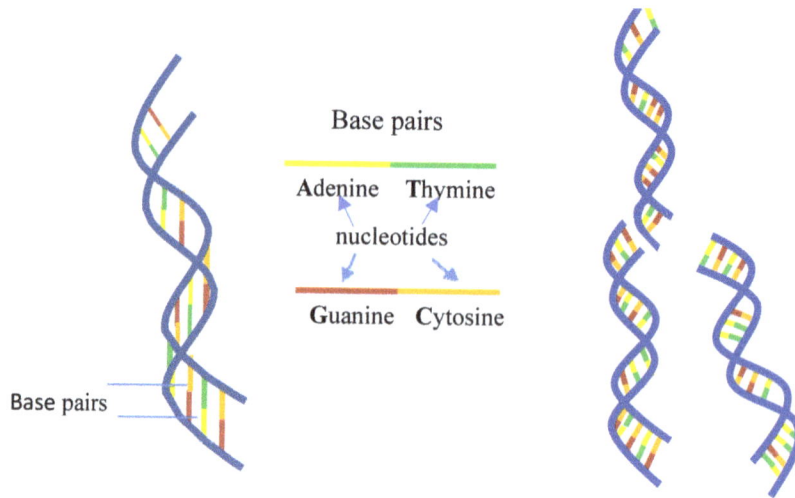

Fig. 6.1 (a) DNA structure   Fig. 6.1 (b) DNA replication process.

Fig. 6.1 DNA.

We saw above that when a cell divides, the DNA divides with it [Fig. 6.1 (b)]. The process of DNA replication is reminiscent of a zip unfastening. The molecule divides lengthwise in two, with each part taking half the base pair rungs of the ladder. These half-base pairs match up with the correct new partner nucleotides produced by the cell – remember A can only match with T and C with G – to recreate the rungs as they existed before, but now in duplicate. The result is an exactly matching replicated cell. Occasionally, errors do occur which can result in diseases such as cancer. On the other hand, these 'errors' can occasionally be beneficial, and become part of the DNA that is passed on to future generations. This is the way that nature evolves.

The origin of life

The division of the DNA was the first key step for life. Single-celled organisms were the first form of life. These cells were of two types: *eukaryotes* (which as we saw above, have clearly defined nucleus; these include animal and plant cells), and *prokaryotes* (which do not have a clearly defined nucleus; these include bacteria). They both multiply by division of their cells into two, but their cells divide in different ways. We will not cover the details of a prokaryotic cell division here. The division of the DNA in eukaryotic cells takes places in the fascinating way we have seen above. However, the result of the division in each case is that every cell split into two cells, each with its own identical copy of the DNA [Fig. 6.1 (b)].

Pretty soon the lakes on Earth abounded with the minuscule creatures that developed because of cell division.

However, these early eukaryotic cells still lacked an essential capability they required if they were to thrive. They needed to be able to produce energy themselves within their own cells. How they managed to do this, is the next fascinating step in the story of how life got going on Earth.

Inside each eukaryotic cell today is another especially important organelle (a small structure which performs specific jobs in the cell, like an organ does in a body) called a *mitochondrion* (plural *mitochondria*) with its own DNA. Mitochondria have the capability of taking in nutrients and breaking these down to produce energy. This was brilliant as all sorts of things became possible once this could be done.

About 1.7 to 2 billion years ago, it is believed that a eukaryotic cell absorbed a mitochondrion within its membrane. The two started to function as partners, the mitochondrion living within the 'mother cell'. This symbiotic partnership was great for both. The mother cell provided the nutrients in the form of oxygen, fats, sugars, and proteins to the mitochondrion. In return the mitochondrion produced energy for the mother cell.

Mitochondria evolved from prokaryotic cells and are part of single-celled life forms which we call 'bacteria'. You may have heard that bacteria can cause us to fall ill. Well, mitochondria are a group in the bacteria family that keeps us alive. It is interesting to think that each of the trillions of cells in our bodies has a microbial (related to microbes, including bacteria) descendent living within it, helping us to grow and develop and function. Not just us, the same is true of every other animal and plant on Earth.

The process of changing light energy into food is called photosynthesis. Photosynthesis is the mechanism that trees, and all other plants use to produce food they need to grow. In plants, photosynthesis uses chlorophyll, which is the substance that makes the leaves green, and carbon dioxide from the air to produce oxygen and the sugars that the plants need to grow. Chlorophyl exists within organelles called *chloroplasts* which are found in plant cells and are responsible for the photosynthesis process in plants.

Respiration is the process which uses oxygen and the stored food molecules to produce energy for cells. Mitochondria are responsible for the respiration process in *both* plants and animals.

But before the mitochondria could produce energy for its mother cell, it needed access to oxygen. How was this achieved?

There is a type of bacteria which also has a plant-like function. This prokaryotic cell uses carbon dioxide, water, and sunlight to produce sugar in the form of glucose, *and oxygen*. Such bacteria are called cyanobacteria (the 'cyan' in the name tells us that they are cyan or blue green in colour).

Now here are a couple of interesting things about cyanobacteria. You should remember from the earlier chapter on the Solar System that when the Earth first formed, its atmosphere contained a lot of carbon dioxide, but no oxygen which is essential for creatures like us to survive. Therefore, at that time, there was no animal life at all. And there never would have been unless most of the carbon dioxide could be changed to oxygen. Well, some 3.5 billion years ago the cyanobacteria came galloping to the rescue. Over millions of years these remarkable organisms slowly converted the atmosphere from one poisonous to advanced life forms such as us, to one that had plenty of oxygen. We owe our existence, and that of other advanced life forms, to the lowly bacteria.

What became of the cyanobacteria? They thrived. They are the oldest life-form and are still around. Those bacteria that died over the ages, sank to the bottom of the seas. Where the seas were shallow, they collected in clumps. Today we find flat, stone-like structures called *stromatolites* in shallow seas around the coasts of some countries in the world, including Shark Bay and nearby Hamelin Pool in Western Australia [Fig. 6.2]. Remarkably, these structures are the remains of the cyanobacteria that were responsible, all those billions of years ago, for transforming the early atmosphere into the breathable one we have today. Living cyanobacteria still exist and can be found in reefs and in aquaria.

Fig. 6.2 Stromatolites, Hamelin Pool, Australia.

Chapter 6. Are we alone…?

There was another source of energy, apart from the Sun, which was available on the young Earth. At the bottom of our seas there are vents, like chimneys, through which water and gases arise from the mantle under the sea floor. The material bubbling out of the 'deep sea vents' is full of chemicals which react with each other to produce other chemicals and release energy [Fig. 6.3]. Recently, scientists have found that hydrocarbons, which are the building blocks of life, are being produced there. Surrounding the tops of these chimneys are found 'worms' and other strange marine creatures thriving by using energy released at the vents.

Some scientists believe that life on Earth may have originated at the mouth of such deep-sea vents, using energy arising from the bowels of the Earth. Of course, it may have originated both ways.

Fig. 6.3 A deep-sea hydrothermal vent (*smoker*).

What really makes it interesting is that we believe that there are *smokers* (as deep-sea vents are also known) under the oceans on some of the moons of the gas giants, such as Europa (a moon of Jupiter) and Enceladus (a moon of Saturn), that could be supporting such primitive forms of life.

The evolution of life

Whichever was the way that life originated, the seas were soon full of these single-celled organisms, dividing and multiplying. Over the billions of years that followed, small natural changes in the DNA in genes resulted in different types of cells. Cells started to stick together in clumps to form multi-cellular life. Other cells evolved and specialised to perform different functions that gave advantage to their life forms. The Cambrian geological period around 540 million years ago was an especially fruitful time for the development of animal species. The first eyes appeared at the beginning of this period. Even though these eyes could barely distinguish light from shade, they allowed their owners to find food, shelter, and escape danger. Soon more complex eyes evolved. Some evolutionary biologists believe that the process from rudimentary eyes to ones such as ours could have taken a mere half a million years.

The life forms became more complex and evolved into many different types. Some were predators, and some the prey. Some were small, and some big. But thus far, all lived in the oceans. Then a creature may have appeared due to random DNA changes which had primitive lungs as well as gills, one that could breathe air as well as take oxygen from water. It may have found itself on dry land, thrown there by waves at the shore of the lake where it lived. It flopped about, then re-entered its watery home. Over generations it was able to stay longer and longer above and outside the water. Its fins slowly evolved to form primitive legs. It could now stagger, rather than simply flounder about on land.

You can see where this is leading. The changes to the life forms are occurring through changes to the DNA, which are themselves the result of chance. Some of these changes give the life form an advantage and are retained. Others that result in a disadvantage are soon lost because those creatures do not survive. The process is called 'natural selection'.

The theory of evolution through natural selection was developed by a particularly important English scientist called Charles Darwin (1809-1882). Darwin took a long trip from 1831 to 1836 in a ship called the Beagle, when he sailed around the world studying the various types and varieties of animals he encountered. He was trying to figure out the reason why these varieties existed and how they came about. He found many birds on isolated islands which had features, such as the shape of their beaks, which helped them eat the type of fruit and seeds found on their island alone. Darwin noticed that these beaks differed from the beaks of the birds of the same species on other islands, which had a different diet. He concluded that the animals had evolved to fit in with their environment. In other words, their differences had evolved to give them an advantage in the environment in which they lived. On his return to England, he wrote an extremely important and influential book called *On the origin of species* to explain his theory of evolution.

We see that over billions of years, millions of species have evolved on Earth to take advantage of the environment in which they found themselves. Some have been tiny, some huge. Some are land-based, some live in the sea and some fly. Some, like the dinosaurs, ruled over the Earth for many hundreds of millions of years, till they were destroyed by a change in the environment with which they could not evolve fast enough to cope. In the case of the dinosaurs, this was due to a huge asteroid hitting the Earth, perhaps exacerbated by the vast volcanic eruptions which were occurring at about the same time. These events completely changed the climate and destroyed the dinosaurs' food sources. In other extinction cases, the cause was an ice age, or other significant climactic changes. Today, sadly, it is the actions of man that are causing the change in the climate, altering the habitat, and killing the animals, resulting in the loss of many species.

Chapter 6. Are we alone...?

What the future will bring is uncertain. What we know is that evolution has resulted ultimately in a species with the intelligence to be able to think for itself, question the reasons behind the environment in which it exists, and to try and do something about it. The intelligent species is, of course, us, homo sapiens. To question is good. That way lies knowledge. But knowledge also brings responsibility. Are we up to the challenge? Time will tell.

However, this story is not about us. It is about the bigger picture. It is about whether we are alone, the sole intelligent beings. Or are there others? But if there are, where are they? How can we find them? Can we make contact?

Where is ET when we need him?

What do we need for life?

We know that there is at least one place in the Universe where life took root and evolved. That of course is Earth. But is Earth unique? That would seem highly far-fetched as there appears little reason for our home to be the only chosen one, when you consider how big just our own galaxy (the Milky Way), let alone the Universe is. It is estimated that there are some 200 billion stars in the Milky Way, and that there are some 200 billion galaxies in our observable Universe. If our Sun is the only star whose solar system has a planet that has life, it would mean the other 40,000,000,000,000,000,000,000 or so stars have not been so lucky. Those odds are just too big for me. In my opinion the likelihood is that there are many solar systems with planets where life exists. But which ones? Finding these is quite another matter.

How should we find the planets where life exists? What should we look for? Well, let us start by defining what we are looking for.

We know that life here on Earth is carbon-based. This means that all living creatures on Earth are primarily formed of molecules made of carbon and hydrogen (also known as hydrocarbons). The reason this is important is that carbon atoms, in combination with hydrogen, the most common element found in the Universe, have the ability of forming exceptionally long molecules. This allows the complex molecules needed for life to be produced relatively easily. An example of such a complex molecule is DNA. It is feasible for other elements such as silicon also to form large molecules, and so to be the basis of life. But to our knowledge, nothing beats the carbon-hydrogen combination. So, let us keep to this.

We also know that for such molecules to form, we need an environment which provides the means for the component elements to come together and chemically react to form other, more complex, structures. The best such environment we know of for this is liquid water. Water can dissolve most other elements and can provide the facility for the elements to be transported around till they meet other elements with which they can chemically react.

Note that we are talking about *liquid* water. Water freezes into ice at 0 C and boils away into steam at 100 C at normal pressures. Therefore, for water to exist as a liquid it needs to be in an environment which is at a temperature between these two limits, and where there is an atmosphere or a cover (such as an ice sheet) to stop the water evaporating away.

For liquid water to exist and to move about freely, it needs to be standing or flowing on a material with which the water itself does not react strongly, and which is solid so that the water does not seep away. Rock is a good example of such a material.

So far for life to exist we have identified the need for carbon and hydrogen, a temperature range of 0 C to 100 C, water, an atmosphere, or other covering, and a rocky, or at least a solid, base.

If we were to search for a planet which meets these criteria, in the first instance, we would look for rocky planets and rule out gaseous ones, such as Jupiter and Saturn, as well as planets where the temperature never drops below 100 C or never rises above 0 C.

We also know that rocky planets exist because there are four major ones in our solar system alone, together with loads of minor planets and asteroids.

In the previous chapter on the Solar System, we talked about the Goldilocks zone of a solar system. This is the part of the solar system which is far enough from its parent star for the temperature to be below 100 C and near enough to the star for its temperature to be above 0 C. We know such zones exist because we live in one.

We also saw in earlier chapters that all the hydrogen that exists in the Universe was formed in the Big Bang. Hydrogen is the most common element and there is plenty in the Universe. All the carbon that exists was formed in the burning cores of stars and distributed across the nearby space when the stars ended their lives in a supernova explosion. A lot of stars have died, and distributed their content elements far and wide, so plenty of carbon exists.

As you can see there is reason for optimism that planets that could potentially host life should be plentiful. Yet, until the late 1990s, no planets at all had been discovered outside our Solar System (these are called exoplanets), and many people doubted that any would be found.

The search for exoplanets

The very first exoplanets were found in 1992. These were four planets orbiting a pulsar. The first exoplanet that was orbiting a normal star was discovered in 1995. This was a giant planet which orbits a star called 51 Pegasi in only four days. 51 Pegasi is in the constellation of Pegasus and is 50.9 light years from the Earth. It is not surprising that this early discovery was of a giant planet, orbiting very rapidly around its parent star, which itself is relatively near to us. To understand why, let us see how astronomers find these far away, tiny worlds.

Chapter 6. Are we alone…?

Planets are extremely small in comparison with their parent stars. For example, we know that well over one million Earths will fit inside our Sun. Further, planets do not shine with their own light, but only by their sun's light, that they reflect. So, if we have a tiny planet going around a huge star, its reflected light will be drowned out by the sun's glare making the planet impossible to see. Since we cannot see a planet directly at present with the instruments we have, the scientists have had to come up with other solutions. These are summarized in Fig. 6.4.

---

(1) Radial Velocity – Watching for wobble

A star wobbles due to the planet's orbit around it. This results in tiny changes of the frequency of the star's light. The resulting subtle changes in the star's colour can be measured.

(2) Transit – Searching for shadows

A planet passing in front of its star dims the star's light which can be measured.

(3) Direct imaging

A picture can be taken of the planet despite the glare of the star.

(4) Gravitational microlensing

Einstein's gravitational lensing technique can be used to measure the bending of the star's light by the planet's gravity.

(5) Astrometry – Minuscule movements

The tiny wobbles caused to the star by the orbiting planet can be measured.

Fig. 6.4 Five ways to search for exoplanets.

One solution is to try and find planets whose orbit is directly between the Earth and the planet's star. [Point (2) Fig. 6.4] If we now carefully measure the light reaching us from the star using a powerful telescope, we find that the light dips when the planet goes across the face of the star. This tells us that the star has an orbiting planet. Of course, it helps if there is a big dip in the light. This occurs if the planet is exceptionally large, and/or if the planet is extremely near its star. The nearer the planet is to its sun, the quicker it will go around the sun. In our Solar System, the Earth takes one year to orbit the Sun, while Mercury which is the planet nearest to the Sun goes around in only about 88 Earth days.

Another way is to use the effect that gravity has on orbiting objects. [Point (1) Fig. 6.4] There is a simple experiment that will explain what I mean. You may have played a game called swingball, (also sometimes called tetherball) where you have a pole stuck in the ground with a ball attached to its top with a rope. If you hit the ball so that it goes around the pole, you will find that the top of the pole itself does a small orbit of its own. Imagine now that the ball you hit is a planet that it is going around the top of the pole which represents the sun. If you look at the 'sun' from where you are standing, you will be seeing it wobble around its centre. This is exactly what happens in real life, where the planet (ball) is going around in an orbit not because of a rope but because of the gravitational force between the planet and the sun. If you examine the sun, which is the star in the sky, very carefully, you will find that it does a wobble if it has one or more planets going around it. The size of the wobble also tells the astronomer how near the planet is to its sun.

The astronomers use these and other techniques to find out how many planets the star has, how big they are and how near they are to their star. They can also tell how massive the planet is, and how dense (in other words, whether the planet is rocky like the Earth, or gaseous, like Jupiter and Saturn).

You see how much we can find out about what is in the Universe simply by looking at the stars in the sky. As we have said before, looking means examining the light that is reaching our eyes from the distant places, which is all we can do.

Up to June 2020, over 4250 planets in more than 3,100 systems had been discovered. Almost 700 of these systems have more than one planet orbiting their sun. The number of exoplanets found is growing almost daily. Look up NASA and other such websites, or query Google, to check the latest count of the exoplanets. The size of the planets we are finding is also getting smaller. Remember we are seeking Earth-like planets which are in their Goldilocks zones. We think we may already have found some. One of the most encouraging of those planets, in constellation Cygnus, is KOI-7923.0 discovered in 2017 by NASA's Kepler Space Telescope. The planet's year lasts 395 days, in comparison with ours of 365 days, and it is just a wee bit smaller being 97 per cent the size of Earth. The sun that the planet orbits is a little cooler than ours and the planet orbits it is a bit further away, making it a little colder than Earth.

Then in January 2020, NASA announced the discovery of the Earth-sized planet, TOI 700d, in its star's habitable zone. NASA's TESS (Transiting Exoplanet Survey Satellite) made the discovery which was confirmed by NASA's Spitzer Space Telescope. NASA's Kepler Space Telescope has also been involved in several other discoveries of potentially interesting candidates.

Chapter 6. Are we alone…?

It appears that stars with planets are two-a-penny, and even those with planets in their Goldilocks zones may be common.

With increasingly powerful telescopes being sent out into space, our knowledge of exoplanets will grow by leaps and bounds. The future looks bright for this part of science. After we have found the right type of exoplanet the right size and distance from its sun, we will check if it has water, and then if there are indications of life. How will we do that?

Well, we will need even more sensitive instruments. First, we must determine whether the planet has an atmosphere. I want you to remember something we discussed in our earlier chapters. We can tell what elements that the light entering our eyes met along its journey from a distant star to us. We can do this because elements absorb light of specific frequencies. You will recall that if we examine the spectrum of the star light, we can look at the black lines (absorption lines) in the spectrum and identify which elements were responsible for them. For example, we can tell that the light met hydrogen along its path if we see the absorption line at the hydrogen frequencies. Now it so happens that we can also identify not just single elements but also complex molecules that the light met. So, for instance, it can identify water, oxygen, carbon dioxide, methane, and other molecules, including organic ones. If our exoplanet has an atmosphere, then when it is directly between us and the star, the light from its star will pass through its atmosphere on its way to us. If our instruments are sensitive enough, they will identify the distinctive components of the planet's atmosphere.

The important molecules we are interested in are water and methane. Water indicates that there is probably water on the surface from where the molecules have evaporated into the atmosphere. Methane indicates life. Life as we know it on Earth produces methane as part of its life cycle. But methane does not last long in the atmosphere; it combines with other molecules to create water vapour and carbon dioxide. Therefore, if we see methane in the atmosphere it means that it is being replenished from the surface. This will be a strong indicator of life. Of course, we will not be able to tell what type of life it is. But just knowing that there is a strong likelihood of life will be a highly significant pointer that we are not alone in the Universe.

We need to find life in just one other location than Earth. Because, once life is found in one other place, we will know that life can evolve independently of Earth and can therefore occur wherever the conditions are right. There are so many, many stars in our Universe, and therefore so many, many planets, that the Universe must be teeming with life. But there is life and then there is life.

The first life on Earth was single-celled; bacteria is such a life form. Today, there are believed to be some 9 million different life form species on Earth. These are scientifically organised in groups ranging from fungi (such as mushrooms) to plants (trees, grasses, mosses) and animals (worms, birds, mammals, fishes, humans) and protista (kelp, some amoeba, mosses).

Quite apart from the desire to find life on a world other than our own, what has fascinated human beings is the possibility of finding *intelligent* life, one with which we could possibly communicate. This possibility has been the source of a lot of science fiction stories. Popular myths and conspiracy theories have grown up, such as those relating to UFOs.

But what does science have to say about this? Is there extraterrestrials life in the Universe (does ET exist)? What are the chances of us finding it? Or it finding us? What are the dangers? We will try and answer these questions. But first let us speak more about life: whether there is life in our own Solar System apart from that on Earth? What are the chances for life elsewhere in the Universe, and how do we look for it?

Extremes of life

We know that life exists in many forms on Earth. We have already talked about life starting as simple, single-celled organisms we know as bacteria. Bacteria are the most common form of life. They are of many types; some are harmful to humans (such bacteria are also called microbes) and some are benign. For this chapter, the interesting thing is that they are found in virtually every environment on Earth, some in which no other life form could survive. These are called extremophiles (or lovers of extremes – the prefix 'extremo-' signifies extreme conditions and the suffix '-phile' signifies love for). We have extremophiles which thrive in dry, acidic, high pressure, radiation, chemical and other extreme conditions. Some bacteria can withstand the vacuum of space and could easily hitch a ride on asteroids. Others have been found living in liquid tar, under miles of ice in Antarctica, deep underground, in boiling waters of geysers, in exceedingly salty seas, and in high radiation conditions.

Because of the versatility of microscopic life, it is extremely important that we do not accidentally transfer life from Earth to another world we are exploring. Either by taking it with us when we go there ourselves or sending a contaminated probe or rover to its surface. This is the reason spacecraft undergo extensive decontamination and sterilisation before they are launched.

The reason why these microscopic life forms are of interest to us is that if we want to find extra-terrestrial life, we must make sure that we do not only look for intelligent life which is probably rare. The most likely form of life we will find is likely to be bacterial. On Earth, all life evolved from the primitive single-celled creatures. Therefore, finding such life on other worlds will be important, even before more complex life forms have been discovered.

Chapter 6. Are we alone…?

The evolution of life on Earth beyond the bacterial stage involved the development of plants, fungi, and animals. We do not know what direction life might have taken on other worlds. On Earth, our life is based on carbon, hydrogen, and oxygen chemistry. Is it the same elsewhere? There are other chemistries possible which allow complex molecules to be built. Will we find that that has occurred on the worlds we explore? We just do not know.

Life in the Solar System

Surprising as it may seem, life in our Solar System, apart from our own world, is feasible. Before we started our exploration of the Solar System in the 1960s, it was unthinkable that there would be any place in our Solar System where there was a possibility of finding life, with the exception perhaps of the planet Mars.

Historically, Mars and Venus had held the possibility of life in our imagination. They figured in our literature, particularly Mars through books such as *The War of the Worlds* by H G Wells, and through erroneous sightings through early telescopes of the planet's so-called canals. Venus was also held as a possible inhabitable world in popular imagination. Modern science soon proved that Venus was a hell rather than a heaven because of its extremes of temperatures and pressures. Mars too proved to be an arid world.

But even for these two planets, all possibilities of present or extinct life have not been totally disproved. There are suggestions that despite its hostile environment, there is a remote possibility that life may exist in the Venusian upper atmospheric clouds where the atmosphere, temperature and pressure are tolerable. Mars also shows exciting possibilities.

<u>Stop press</u>: It was announced on 14 September 2020 that a gas called phosphine has been discovered in unexpectedly large quantities in the upper reaches (more than 55km above the surface) of the Venusian atmosphere. The excitement is because phosphine is associated with microbial life. It disappears rapidly and needs to be replenished. The question is how this is happening. So far, all known explanations of ways to produce phosphine by a non-living source have not been enough to generate the volume of the gas that has been found. The situation is not proof-of-life, but it does raise some interesting possibilities. Keep an eye on this one.

There is water on Mars. It is frozen as ice but is known occasionally to flow. There certainly was liquid water in the past. Perhaps therefore, there is bacterial life under the Martian ground, or at least evidence of life that had existed long ago in the past. This search is ongoing. The rover Curiosity wandered around on the planet sampling the soil. In June 2019 it was reported that Curiosity had discovered methane, which is a by-product of life. This does not mean that life exists. The gas could have been produced billions of years ago, and then got trapped under the surface till it was released, or the methane could have been produced by geological processes. In any case the signs are encouraging, particularly as organic compounds have also been found on the planet's surface. A human mission is being planned over the next 30 years which could prove things one way or the other.

Beyond Mars and Venus some exciting new possibilities for life have been identified, in regions that have surprised all – the cold moons of Jupiter and Saturn.

Jupiter and Saturn have been visited by several flyby probes, including Voyagers 1 and 2. These spacecrafts photographed most of the major moons of the two gas giants and showed them to be frozen, icy planets. The exception was Titan, the huge moon of Saturn, which showed itself to be covered by a thick impenetrable orange haze. Two of the frozen moons of Jupiter also excited the astronomers' interest: Ganymede and Europa.

Ganymede [Fig. 5.21] is Jupiter's largest moon and is also the largest moon in the Solar System. It has a radius of 2,630 km and is only a little smaller than Mars whose radius is just under 3,400 km. Ganymede is completely covered in ice but is believed to have an ocean of salty water about 150 km deep, some 100 km beneath the surface. How do we know this? Well, we know that, remarkably, uniquely among the Solar System moons, Ganymede has a magnetosphere. What this means is that Ganymede has a magnetic field, just like Earth, Jupiter, Saturn, Uranus, and Neptune in the Solar System. For a body to have a magnetic field, part of its core must be liquid churning around a solid core. In the Earth's case the inner core is solid iron, and the outer core is liquid iron. In the case of the gas giants, the inner core is probably rocky or solid hydrogen with liquid hydrogen spinning around it. Ganymede is thought to have a rocky inner core with a salty water outer core. We know that the outer core is probably salty water because salty water in motion around a rocky core can generate a magnetic field. We can also work out how deep and big the ocean is under the surface by measuring the density of the moon and knowing the density of salty water.

The final proof of Ganymede's magnetic field came when aurora lights were seen around the moon's poles by the Galileo spacecraft on 28 December 2000. These aurorae were just like the aurora borealis (northern lights) and aurora australis (southern lights) around the Earth's north and south poles, and the auroras around the gas giants' poles. You will remember that auroras are caused by the solar charged particles being directed to the poles by the planet's magnetic field.

But why is an ocean so exciting in our search for life? Remember we said that the hydro-thermal chimneys at the bottom of the Earth's oceans were found to be teeming with life. Scientists believe that such chimneys may have been where life began on Earth. If so, life may have originated deep under Ganymede's seas too.

Chapter 6. Are we alone…?

The second Jovian moon to excite the astronomers was Europa [Fig. 5.21]. Photographs show that the frozen surface is icy but complex. It looked as if there were icy slabs floating about, sometimes crashing into each other in jumbled pack ice, and sometimes drifting apart, leaving smooth, icy channels between them. The conclusion was that there indeed was ice covering the whole moon, and, excitingly, the ice was not smooth and solid. This could only happen if the surface were thawing into huge icy sheets, moving around and then refreezing. The excitement was that for this to happen there must be a source of heat that caused the ice to melt, and that the surface ice was floating about on liquid water which came up to the surface as the ice sheets moved apart and then refroze. The source of heat was soon identified as the tidal effect of Jupiter and the other moons of this planet tugging on Europa, more as its orbit brought it nearer to the planet and then less as its orbit took it further away. This is rather like you squeezing and releasing a tennis ball which soon heats up, which we mentioned earlier. In Chapter 5, we mentioned a possible NASA project to explore Europa's oceans.

But why is this of interest for our search for extraterrestrial life? You will remember that our key criteria for worlds where life could exist was liquid water. Well, so far, we have noted three places in our Solar System apart from our own home Earth, where we believe liquid water exists: Mars, Ganymede, and Europa. Amazingly, they are not the only ones. There is another body where we know that there is liquid water. This is even further away from the Sun than Mars and Europa. It is tiny Enceladus, a moon of Saturn which is only 504 km in diameter which orbits its parent planet in less than one and a half days. We know that it has liquid water because we have photographed it and analysed it, so we even know what the water contains. But how did we do it?

In 1997 NASA sent a spacecraft, Cassini, on a long journey to Saturn. It turned out to be one of the most successful space missions we have had to-date (2020). It reached Saturn and went into orbit around it in 2004. Over the next 13 years, Cassini visited the rings and various moons of Saturn, photographing and analysing them using the instruments on board. Cassini transmitted a wealth of information back to us on Earth, till on 15 September 2017 at 11:55 GMT, its fuel gone, it was made to fall into the atmosphere of Saturn and burn up.

You should search for Cassini mission online, to read about its achievements and to see the stunning images from the voyage. Here we will talk about what it found on Titan and Enceladus, two of the 62 moons of Saturn that we know about at present.

As Cassini passed by Enceladus on its orbits, it noted that the moon had an atmosphere which the analysis showed was mainly made up of water vapour. Scientists believe that Enceladus is heated by gravitational tides (as is Europa, the Jovian moon we met earlier) and beneath its ice cover lies a vast ocean of liquid water. This was exciting enough but then later huge geysers were spotted rising hundreds of kilometres in the South Polar region of the moon [Fig. 5.23 (b)]. In October 2015, Cassini flew less than 50 km (30 miles) above the moon's surface and through the water plumes. They discovered that the plumes were indeed water and found carbon dioxide and complex hydrocarbon molecules in the spray.

This raises the possibility that in the rocky core of the planet below the ocean there may be hot vents, or smokers, as there are on Earth [Fig. 6.3]. Remember there is a serious theory that life could have originated on Earth at such vents. All these exciting findings makes Enceladus one of the most serious contenders for having microbial life in the sea under its icy cover. Watch out for future missions to this tiny, but especially important moon.

The other moon of Saturn, which is of interest to us from the point of view of life, is Titan. Titan is at the other end of the scale in size to Enceladus, being the second biggest moon in the Solar System, and almost as big as Mars. We have described Titan in the previous chapter on the Solar System. It is believed to be what the Earth was like soon after it formed, having a methane atmosphere and hydrocarbon lakes and seas. In a few billion years when the Sun is dying, and it bloats up and the Earth is swallowed up or burnt to a cinder, Titan may come into its own, warming up and going through a regeneration rather like our home planet did.

But there is another intriguing possibility. A methane-based chemistry may already be operating on Titan. This could lead to a life form quite different from our own carbon-hydrogen based life. Has this happened on Titan? Will a future mission to this large moon turn up such a truly alien life form? Unlikely. We will not know for certain till we search for it.

Meanwhile, the question remains: what is the likelihood that *intelligent* life exists in the Universe?

The Drake Equation

It would be very surprising if we are the only intelligent life ever to have existed in the vastness of our amazing Universe. But can we put a number on the possibility that there is an *ET* out there, somewhere? Well, one US astronomer decided to try. His name is Frank Drake (1930-). He was involved in setting up the SETI (Search for Extraterrestrial Intelligence) Institute and still works on the project as an Emeritus Trustee. We will learn more about SETI shortly. First though, let us see how Drake tackled the problem of figuring out the chance for intelligent life in the Universe, and see what this turns out to be.

Drake developed an equation, which is called the Drake Equation after him, to enable us to calculate this chance. But rather than immediately dive into calculating an equation, let us understand how Drake went about his task.

Chapter 6. Are we alone…?

Drake decided to search for the intelligent civilisations with whom we could communicate. Note that this is not the same as the number of planets on which life has evolved. Rather than search the whole Universe, Drake sensibly limited his scope to our Milky Way galaxy, which you will recall is vast enough with an estimated 200 billion stars. Beyond that, communication will be virtually impossible due to the distances involved. Mind you, this is based on knowledge of science that we have at present. Who knows what may be discovered in the future, and where advances in science may take us? But we will always be limited by the laws of physics, such as the speed of light, which Einstein showed is the fastest anything, including information, can possibly travel.

Drake put down all the things that would affect the possibility of life arising in the Milky Way. This way he could see how often intelligent civilisations are likely to be created in our galaxy, and how long they will be around to enable us to communicate with them. We know that the answer must show that there is at least one intelligent civilisation that has arisen, since we are here. But what are the factors that we should consider when calculating how many intelligent civilisations apart from ourselves are likely to be around. Frank Drake came up with seven factors. Let us list them down:

1. The rate of star formation per year in our galaxy.
    This shows how many possible solar systems are being created every year in the Milky Way. This is the right starting point because the more the stars that are created every year, the greater will be the number of intelligent life forms that can arise each year. We know that our home the Milky Way galaxy is almost as old as the Universe. The number of stars it possesses has been estimated as 250 ± 150 billion, or between 400 and 100 billion. Let us assume that the number is some 200 billion (200,000,000,000) stars. Let us start with that number of stars having formed in the 13.8 billion (13,800,000,000) years since the Big Bang. That gives a rate of about 14 new stars per year.

2. The percentage of stars which have planets.
    Recent work shows that a large proportion of the stars have planets. Let us assume that 1 out of every 5 stars (20%) have planets.

3. The average number of inhabitable planets in the stars that have planets.
    Let us say that there is 1 inhabitable planet possible for each star that has planets circling around it.

4. Percentage of these planets where life develops.
    Let us assume that 1 in 10 (10%) of these planets develop life.

5. <u>Percentage of the planets where life starts, and which goes on to develop intelligence.</u>
Let us consider that 2% (1 in 50) of the planets where life develops succeed in developing intelligence.

6. <u>Percentage of the intelligent life that goes on to develop technology to communicate across space.</u>
Let us generously consider that 5% (1 in 20) of all intelligent life develops such technology.

7. <u>The length of time that such civilisations exist.</u>
Let us be optimistic and say that such civilisations last for 500,000 years.

Let us now put all this together:

Number of civilisations that exist in the Milky Way and who can communicate across space is = 14 x 20% x 1 x 10% x 2% x 5% x 500,000 = 140.

Drake's own estimates gave a minimum number of 20, and a probable range of between 1,000 and 100,000,000. You can have a go yourself.

In June 2020, the Astrophysical Journal reported research done at The University of Nottingham which came up with a likely figure of 36 civilisations capable of communication currently in the Milky Way. Their average distance from us was 17,000 light years, and the likely numbers ranged from 41 to 211.

If you take a pessimistic view, the number is just 1 (us). If you take an optimistic view, the number could run into tens of thousands.

If we gave up on the possibility of communication, then over the Universe, the number of intelligent civilisations may be as high as 10 million billion (!).

Approximately, of course.

The search for extraterrestrial intelligent life (SETI)

Human beings have wondered for many centuries whether there is intelligent life on worlds other than our Earth. They thought initially only about the planets in our Solar System, the only worlds they knew about. The other objects they could see in the sky were of course stars. But they wondered about two planets in particular: Venus, with its brilliant light, they thought must be a paradise; and Mars, the red planet, they thought must be inhabited perhaps by an advanced, alien warrior race. The initial sightings of Mars's surface through poor telescopes appeared to show lines on the surface, which they took to be canals that had been constructed by the inhabitants to bring water to the plains from the ice of the poles. A famous book *The War of the Worlds* by H G Wells in 1897, was about Martians invading the Earth – they were eventually defeated, if you must know, by the tiny bacteria against which they had no immunity.

Chapter 6. Are we alone…?

Alas, both these ideas turned out to be false. We found Venus was nearer to hell than paradise, with temperatures high enough to melt lead, and pressures more than enough to crush any known life form. Mars turned out to be a cold and barren desert, with barely any atmosphere. Still our fascination with life, and more specifically with intelligent life in other worlds, has not left us. The more we have understood our Universe, and got to know how big it is, and how many stars it has, the more it seems inconceivable that we are the only world where life has taken root. Our search for intelligent life has therefore continued. We have put together many scientific projects to answer this question and spent many billions on the search. So far to no avail. But we continue to look, and listen, and send messages for someone to pick up, and answer. There has been no call from ET yet. Or has there?

Remember that we are talking here about intelligent life. This is different to our search for *any* life which, at the least, we would dearly love to discover. Of course, life needs to have started before it can develop intelligence. This search for (any) life is also underway with a mission to Mars well into its planning phase, and others to the moons of Jupiter and Saturn under consideration.

Our search for intelligent life implies that intelligent civilisations exist on other worlds orbiting around other stars and that they can communicate with us over the vast distances of space. We know of only one way that inter-stellar (between stars) communications can take place, and that is by using electromagnetic waves. By now, we are well familiar with these. You will recall these light waves cover the whole spectrum from microwaves with their large wavelengths, through radio waves, to (for us) visible light, and then onto the extremely high energy and short wavelengths of gamma rays. This is the fastest means of communications we know, since nothing can travel faster than light. You should also recall that the distance that light can travel in a year at its speed of 300,000 kilometres (187,500 miles) per second is called a light year and is about 9.5×10^{12} km or 5.9×10^{12} miles.

Now here is the problem. The nearest star to our Sun, Proxima Centauri, is just over 4 light years away and its light takes more than 4 years to reach us. So, if there was an intelligent civilisation on a planet around this star, it would take over 4 years for our signal to reach them and, if they reply immediately, another 4 years or so for their reply to get back to us. Some 9 years per exchange is no way to have a meaningful conversation. Further, to have a reasonable chance of finding an intelligent civilisation, we will need to search many solar systems. Therefore, we should be searching planets around stars that lie many hundreds of light-years from us. I hope you see the problem. How do we make contact? How do we communicate?

The first thing we do is listen. If intelligent beings exist, perhaps they have been communicating already and their signals have already reached Earth. Perhaps they are highly evolved and acquired their technical knowledge of radio communications many thousands of years ago. On Earth, we have been able to send radio signals only since Guglielmo Marconi (1874-1937), the Italian inventor and electrical engineer, sent the first transatlantic broadcast in 1901.

The weak signals from Marconi's broadcast would have spread 119 light years at the time of my writing. Since then, we have progressed rapidly, through to more powerful radio and TV transmissions, telephone, radar, and other forms of electromagnetic-wave-based communications. We are already planning an inter-planetary internet for when we undertake such trips. But overall, the volume of space we have covered is tiny, a sphere of radius of no more than 119 light years. So, the first thing we must do, as we said, is to listen for any communications coming our way.

The story of our listening projects began well over a hundred years ago. As early as 1896, Nikola Tesla (1856-1943), a Serbian/later-American scientist famous for his work on electrical currents, transmission, motors, and magnetic fields, suggested that using his transmitters, we could make contact with beings from Mars. Three years later during his experiments, he believed that he had heard signals coming from that planet. Subsequent analysis showed that he had not picked up any such signal. However, the thought remained that Martians were there and possibly transmitting to us. Then in 1924, when Mars was in one of its closest approaches to Earth, a "national Radio Silence Day" was observed in the United States. For three full days from 21st to 23rd August, all radios were silenced on the hour every hour. The object was to maintain absolute radio silence so that there was no noise generated to disturb the listening radio equipment. Nothing was heard.

Later in 1960, Project Ozma was set up at Columbia University by astronomer Frank Drake (the same one who devised the famous equation we spoke about earlier). Here, Drake used a large radio-telescope (a telescope designed to receive signals at radio frequencies, rather than at visible light frequencies) tuned to receive signals at a frequency of 1.420 Gigahertz. Gigahertz is measure of frequency of electromagnetic (light) signals and stands for 10^9 cycles per second. If you remember our discussions at the start of the book, 10^9 represents 1,000,000,000 or a thousand million (a billion). Thus, at the chosen frequency, the light signal oscillates a thousand, four hundred and twenty million times per second.

There was a good reason for the choice of this frequency: it is the frequency of light given off by hydrogen (H_2) atoms and corresponds to the 21 cm wavelength of the hydrogen spectral line on the electromagnetic spectrum. Hydrogen atoms are the most common in the Universe and the frequency would be a good choice for anyone wanting to communicate with another intelligent species. Drake searched two stars called Tau Ceti and Epsilon Eridani. However, again, no signal was picked up.

Chapter 6. Are we alone…?

Project Ozma was expanded with increasingly more powerful telescopes listening out into space and using increasingly refined equipment to filter out noise from any incoming signals that may be received. Still, the silence was deafening.

Till 15 August 1971.

When astronomer Jerry Ellman, who was working at the Ohio State University Big Ear radio telescope, examined his data, he found the recording of a powerful signal that the telescope had picked up a few days earlier on 15th August, which had lasted a full 72 seconds during which the telescope was able to observe it.

Ellman was so excited that he circled the data on the printout and wrote "Wow!" by the side [Fig. 6.5]. Since then, the signal has been known as the 'Wow! signal'.

Sadly, all attempts to get the signal back, despite knowing the direction in the Sagittarius constellation it came from, have failed. Even now the Wow! signal remains unexplained, and it remains the best (and only) candidate for an alien transmission we have to-date (but see Stop press, below).

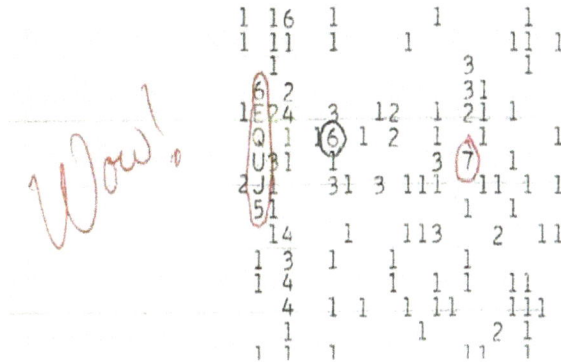

Fig. 6.5 The 'Wow' signal.

On 17 May 1999, University of California, Berkeley launched an initiative called SETI@home. The idea is that anyone with a computer can connect into the project by downloading a piece of software to analyse data received from the telescopes. The participating computer is sent chunks of data to analyse by the University. The software program on the computer takes advantage of any time that the computer is idle to process the data it has received; the results are then sent back via the internet to the University, which collates all the data coming in. The participating computer then receives further work, and so the process continues.

As of March 2020, 1,803,163 users were participating. It is an exciting venture and who knows, the project may discover ET after all. If you are enthused by this, look up SETI@home on the internet, find out about the project and see if you would like to participate.

Caution: *Remember that there may be a cost to remaining connected to the internet and in running the program. So, if you are not the one who pays the charges, you must always check with whoever does, and make sure it is all right with them before you download or signup to anything. You must also be cautious about downloading software from external sources to avoid importing a virus. Always ensure you have good, up-to-date virus checking software on your laptop or PC system.*

The search has continued over the years. The telescopes have grown bigger and more powerful and sensitive. The SETI Institute has collaborated with the University of California, Berkeley to set up the Allen Telescope Array (ATA). ATA opened with 42 radio dish telescopes in 2007; the aim is, ultimately, to have 350 or more telescopes. ATA is named after Paul Allen (1953-2018), the co-founder of Microsoft, whose donation made it possible.

In 2015, another major initiative, "Breakthrough Listen", was announced by Stephen Hawking (1942-2018), the British scientist, and the project's benefactor the Russian billionaire Yuri Milner (1961-). Breakthrough Listen is funded by some US$100 million. The project will search ten times more area of the sky, use a five times wider frequency spectrum for the signals and have a capability to process data a hundred times faster than has been achieved by any previous project. The plan is that the search will cover not only our Milky Way galaxy, but over 100 of the nearest galaxies. It will also be co-ordinated with the SETI@home, so perhaps even you will be able to take part in the search too.

We are certainly not giving up yet.

Stop press: The UK *Guardian* newspaper reported on 18 December 2020 that astronomers at the Parkes telescope in Australia have been working on the analysis of a signal that the telescope picked up in April/May 2019. The 980MHz [980 x 10^6 cycles/second] signal was from the direction of the 4.2 light years away Proxima Centauri, the star nearest to our Sun, and was said to have "an apparent shift in its frequency consistent with the movement of a planet". The scientists "have yet to identify a terrestrial culprit such as ground-based equipment or a passing satellite". While cautioning that the "signal" is likely to have a mundane explanation, the newspaper said that "Scientists are now preparing a paper on the beam, for Breakthrough Listen, the project to search for evidence of life in space". It quotes its source as saying, "It is the first serious candidate since the 'Wow! signal'".

Note: The Breakthrough Listen project mentioned is the one set up in 2015 by Stephen Hawking and Yuri Milner and discussed above.

Chapter 6. Are we alone…?

Concerns and doubts

However, exciting as the prospect of finding intelligent alien life is, there are some concerns and doubts.

The Italian physicist Enrico Fermi (1901-1954) has been famously quoted as asking "Where are they?" in the 1950s referring to our lack of contact with aliens. If these civilisations were as common in the Universe and as technologically advanced as was supposed, we should surely have found them, or they us. This lack of contact has been called "the Great Silence". Perhaps there are not as many such civilisations as we think. Maybe our technology and search strategy are not yet up to the task of making contact. Or it could be that we simply have not searched long enough. The vastness of space and the number of possible locations to search is enormous.

We have only been technically capable of mounting serious searches for well under fifty years. As we have seen, we have been broadcasting our presence through our radio and television signals for about a hundred years. Even if the aliens picked up our weak signals and decoded them and worked out where they were coming from and responded with a powerful signal beamed straight at us, it would take anything up to a hundred years for their reply to reach us.

Perhaps there simply has not been time enough yet.

It is also possible that many of the technologically advanced civilisations just do not last awfully long. Sadly, many may not be able to handle the dangerous technology they have discovered. A nuclear war is more than capable of sending a civilisation back to the stone age. This is one of the dangers also facing our own civilisation, of course. In using Drake's equation, we had glibly assumed that such advanced civilisations will last for 500,000 years on average. What if they lasted just 50,000, or less?

On a more cheerful note, let us assume that there are civilisations around that have figured out how to manage their affairs and to continue to grow and thrive. Such civilisations may have developed technologies far ahead of ours. Their means of communicating may be beyond our knowledge and understanding. In such a case, it is possible that we are not in contact because we simply do not have a common means to do so. To cover this problem, scientists have thought of techniques that can be used. Mathematics plays a big role in this, since it is universally applicable. For example, the ratio of the circumference of a circle to its diameter (we call it 'pi') is a universal constant, that is, its value is the same everywhere in the Universe. There are also other things which should be understandable by all advanced intelligent beings. Examples from science include: the structure of atoms, the frequency radiated by Hydrogen atoms, etc.

The space probes Pioneer 10 (1972) and Pioneer 11 (1973) carry plaques engraved with symbols including human figures, scientific notations and astronomical depictions of the Sun's location and the Solar System's structure with the Earth highlighted, in an attempt at communicating with any extraterrestrials who happen to find and examine the probe. The Voyager 1 spacecraft launched in 1977 carries a golden record which contains sights and sounds of Earth, including messages, music, and scientific information such as the structure of DNA, which is found in all living cells on Earth.

There is the final possibility for our lack of contact, and that is that perhaps these civilisations just do not wish to communicate with us. This is not as strange as it sounds. Here on Earth, we have had distinguished scientists, including Stephen Hawking, caution that we must be careful about contacting an advanced extraterrestrial civilisation. The reason is that there is no certainty that the visitors will be friendly to us humans. Historically here on Earth, the interaction between the advanced and the less developed nations has hardly been a ringing endorsement for such contacts.

Other scientists disagree with the cautious approach. Despite the unease, messages have been transmitted from Earth, directed out to space. In 1974, a message was sent from the Arecibo Observatory towards the M13 globular cluster 25,000 light-years from Earth. Given the distance this was a symbolic gesture; we will need to wait 50,000 years for a reply. Other messages have been sent since. Active SETI, or messaging to extraterrestrial intelligence (METI), is being seriously studied and various checks and tests have been proposed which should precede any such transmission.

So where does this leave us?

The Voyager 1 and 2 spacecrafts are now both beyond the *heliopause*, the boundary where the hot, solar wind of the *heliosphere* - the region of space surrounding the Sun that is under the influence of the solar wind - meets the cold matter and radiation of interstellar space [Fig. 6.6]. They have escaped the influence of the Sun's wind. Voyager 2 is now (January 2021) some 17.5 light hours (18.7 billion km or 11.6 billion miles) from Earth; it responded correctly to a recent contact, the communication was received after 34 hours and 48 minutes of sending. You may recall that the distance of the Earth from the Sun is only 8 light minutes, so Voyager 2 is over 130 times as far. Voyager 1 is even further ahead. It is still able to respond to commands to send images. Will these spacecrafts or their signals ever be picked up by other beings?

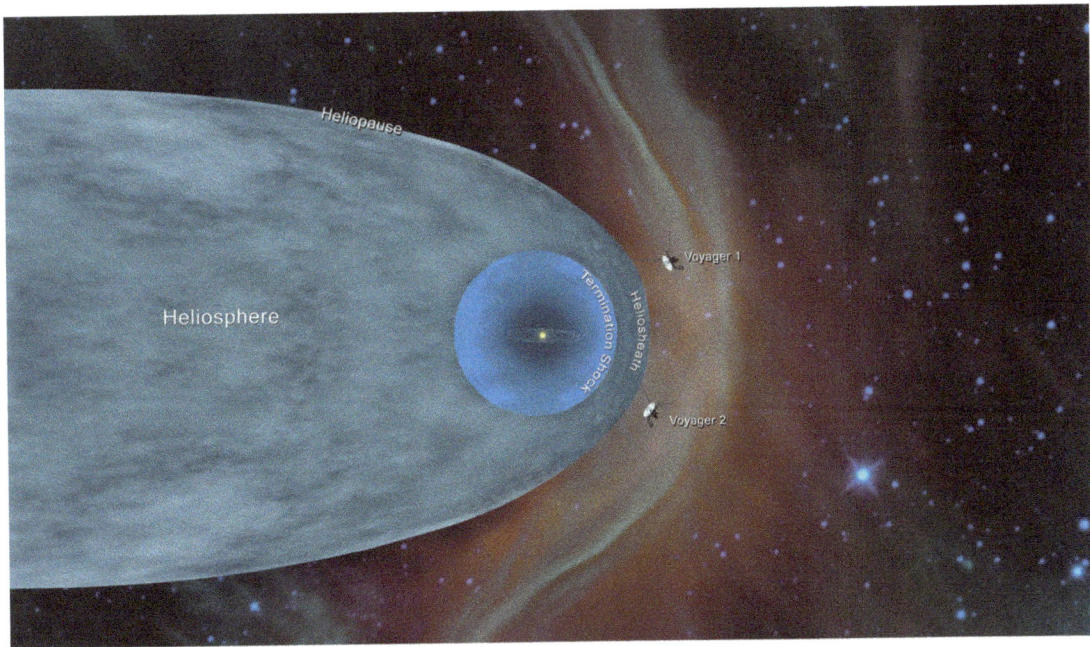

Fig. 6.6 Voyagers 1 and 2 are now in outer space, beyond the heliosphere.

It is certain that any proven contact with an extraterrestrial civilisation will fundamentally change humanity. It could be of significant benefit, or a disaster. But the need to know is a basic driver of our being human. So, we will listen, intently. Whether we will, or should, actively broadcast our presence to the far reaches of space, needs more debate. We have little choice about near space, of course, the cat is out of the bag with the ever-increasing ripples of our radio and TV transmissions spreading to ever more distant worlds. Our first public radio broadcast was made on January 10, 1910, from the Metropolitan Opera House in New York. In January 2010, the signals would have reached 100 light years away. In January 2030 they will be 120 light-years away.

Will any being be listening?

Chapter 7

Farewell to the Universe...

how it may all end

We know that our Universe has been around for some 13.8 billion years. It started in a fiery burst of energy from nothing. The tiny baby Universe expanded at an enormous rate. It grew and cooled. Some of its energy converted into matter, the rest remained as energy in the form of light. The matter collected and formed stars. The stars gathered into galaxies. Still the Universe expanded.

Then what? Will it keep expanding forever? If not, what will stop it expanding?

This chapter examines what is the likely fate of our Universe and how it may occur. We will look at some of the possibilities, and examine which is the most likely, and why. What I can tell you is that whatever is the likely end fate of our Universe, it is a long, long time in the future.

So, let us relax and let our minds wander.

Before we finally close the story, we will also look at some other ideas that scientists are working on, such as: is ours the only universe, or do we live in a multiverse where our universe is just one among many others? Where do these other universes exist if they do? Is there any other way that a big bang could have happened? Do we have any way of finding these things out, or will they remain just fantastic possibilities for ever?

The expanding Universe

We saw in Chapter 1 that Edwin Hubble was able to show that the further a galaxy was from us, the more the wavelength of its light was stretched and the redder the light appeared. Hubble used this 'redshift' to show that this meant that the further the object was from us, the faster it was moving away from us. This could only happen if the whole Universe were expanding, so that every point was moving away from us. We were able to explain this by using the example of the Universe as a balloon with the galaxies as marks on its surface. If we blow into the balloon so that it expands, we find that all the marks on the balloon move away from each other.

Ever since Hubble's work in 1929, the belief had been that the expansion of the Universe had been kick-started by the Big Bang, but that over time, this rate would slow because the gravitational pull of all the matter that exists would overwhelm the initial push of the Big Bang. In other words, the expectation was that the Universe would simply run out of steam with time.

The scientists were therefore attempting to measure the gravitational pull of all the (normal and dark) matter that exists. They calculated an important value called the *density parameter*, written as Ω (the Greek letter omega) which depended on the average density of matter and energy in the Universe.

If the 'inward' pull of the matter does exceed the push the Universe received at the Big Bang, that is if the *density parameter* was greater than 1 (or, as scientists say, if $\Omega > 1$), then at some point the expansion of the Universe should come to a halt and go into reverse. The Universe would then start to contract, speeding up as it grows smaller. Space itself will be contracting. Since everything in space will then be rushing towards each other, the light waves would get squashed together and their frequency will increase. We would have a blueshift instead of a redshift. The result would be that the galaxies would be rushing towards each other. The spots on our balloon will be getting closer and closer. Our balloon would be deflating!

The end would be a spectacular crash, a *Big Crunch* as it has been called. Such a universe is called a *Closed Universe*. The big crunch will occur if the amount of matter in the Universe is more than a critical value needed to overcome the force that has been pushing the Universe to expand.

What would happen after the big crunch? We do not know. Would the Universe be squeezed into a point, as it appeared at the Big Bang, and disappear? Would this be the end of space and time which came into being at the Big Bang? Some scientists have said that the Big Crunch would result in the Universe coming to a halt and then bouncing back in another Big Bang. This has been called the *Big Bounce*. This hypothesis would mean that the Big Bang is followed by an inflationary expansion, followed by a halt, then a deflation, leading to a Big Crunch, resulting in a Big Bang, on and on forever, like a perpetual yoyo.

Chapter 7. Farewell to the Universe…

If there is less than the critical level of matter needed to halt the Universe's expansion, that is the *density parameter* Ω is less than 1 ($\Omega < 1$), the Universe would keep expanding forever at an ever-increasing rate. This Universe is called an *Open Universe*.

There is, of course, another possibility and that is that the amount of matter in the Open Universe is exactly at the *critical* level to bring the Universe to a gradual halt and then keep it there. This is the case where the *density parameter* is exactly 1 ($\Omega = 1$). It will take infinitely long for such a universe to come to a halt, so that by the time the expansion stops all matter in it will be infinitely far from all other matter. Such a Universe is called a *Flat Universe*.

Space in a closed universe closes in on itself, such as the curvature of the surface of a sphere. The three angles of a triangle drawn in such a universe will add up to more than 180° and parallel lines will ultimately meet as do the lines of longitude on a sphere. For an open universe, where curvature of space is analogous to a saddle, the angles of a triangle will add up to less than 180° and parallel lines will diverge. In a flat, critical universe the angles of a triangle will add up to precisely 180° and parallel lines will remain parallel. Note that the triangles we are talking about need to be exceedingly large, measured using stars across our galaxy, since at a human scale, the differences between the sums of the angles of a triangle, whatever the geometry of the Universe, are too small to be used to determine its curvature [Fig. 7.1].

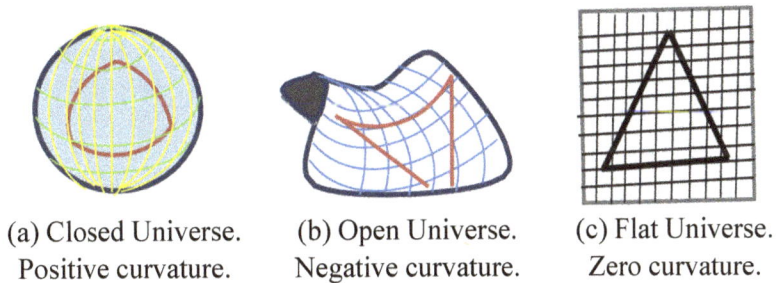

| (a) Closed Universe. | (b) Open Universe. | (c) Flat Universe. |
| Positive curvature. | Negative curvature. | Zero curvature. |

Fig. 7.1 Geometry of the Universe.

Astrophysicists think that we are living in a near-critical Flat Universe. Measurements by WMAP, the Wilkinson Microwave Anisotropy Probe (see Chapter 2), has confirmed that the Universe is flat to within 0.4% of error.

Astrophysicists used to believe that the expansion of the Universe must be slowing down because gravity due to matter in the Universe would be pulling back against the expansion. They wanted to measure how much it is slowing down. Using Hubble's work on the redshift, astronomers were able to work out how far a distant galaxy is from us. Remember the redshift occurs because the light from the distant galaxy is stretched as the Universe expands while the light is travelling towards us. To check the rate of expansion of the Universe, they needed to find out the distances of several distant objects from us measured by another means, which did not depend on the expansion. The distances can then be compared and used to find out whether the rate of expansion has changed.

As we have noted several times before, all that we have available to us to examine a far object in space is the light that is coming to us from the object. One way we can tell how far an object is from us, is by seeing how bright it appears to us. If we imagine a car coming down the road towards us in the night with its lights full on, the lights will be faint when the car is far away, gradually getting brighter as the car comes nearer and nearer. Now if we knew how bright the lights of the car are, and if we knew how bright they appear to be at a distance, we can work out how far the car is at any time. We can do this because of a simple relationship between brightness and distance.

If we have two equally bright objects, one of which is *twice* (2 times) as far as the other, the nearer object will appear *four* (4) times as bright as the farther one. If one is *three* (3) times as far as the other, the nearer one will be *nine* (9) times as bright. Note that 4 is 2 x 2 (the square of 2), and 9 is 3 x 3 (the square of 3). We see that the brightness of an object decreases as the square of its distance from us; mathematically this is stated as the *inverse square law*. Thus, if we measure the brightness of two equally bright objects and object A appears 4 times as bright as object B, we know that Object B is twice as far away from us as Object A [Fig. 7.2]. Similarly, if Object A is 9 times as bright as Object C, we know that Object C is three times as far from us Object A.

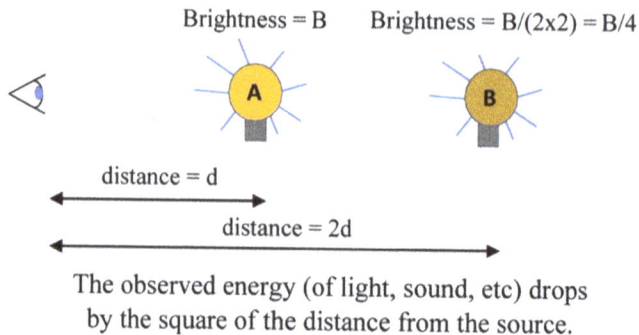

The observed energy (of light, sound, etc) drops
by the square of the distance from the source.

Fig. 7.2 Inverse square law.

We can also draw a graph showing how the brightness varies depending on how far objects are from each other [Fig. 7.3].

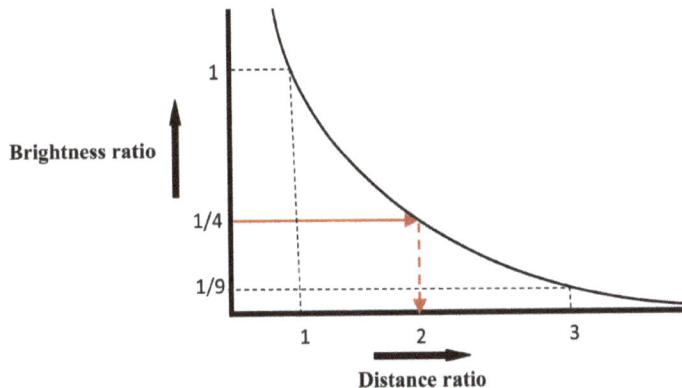

Fig. 7.3 How the brightness of objects varies with distance: the inverse square law.

We can measure how bright a far object *appears* to us, and if we knew how bright it is, we can work out how far it is from us. The astronomers could easily measure how bright an object *appears* to them by simply looking at it. But knowing how bright it was at its source, was trickier. What they needed was a *standard candle*.

A standard candle is an object whose brightness we know, and which will always be the same. Luckily, nature has provided us with standard a candle in space. In Chapter 3, we saw that stars called white dwarfs, which reach the *Chandrasekhar limit* of 1.4 times the mass of our Sun, explode in what is called a Type Ia supernova. This limit is always the same. Since we know the mass, we can work out the energy of the explosion and therefore its brightness.

Finding a Type Ia supernova in a galaxy whose distance we know, allows us to calibrate the Brightness v Distance ratio. Here, then, is our standard candle in space. If we find a Type Ia supernova exploding in a galaxy, and measure how bright it appears to us, we can calculate how far it and its galaxy really are from us.

The problem has changed: we now simply need to find a distant Type Ia supernova in a galaxy whose distance we are trying to measure. But since we do not know when a Type Ia supernova is going to explode, how can we do this?

Luckily, we have captured data on many Type Ia supernovae. This is because while these supernovae in any one galaxy are rare, there are so many galaxies in the observable Universe that a supernova explosion somewhere within viewing range is not that rare.

Having obtained the data, the astronomers drew a graph [Fig. 7.3] of how the measured brightness compared with the known distances to the galaxies in which these supernovae occur. Now astronomers could look at any Type Ia supernova in any galaxy, measure its brightness and read-off its distance from us from the chart.

If the Universe were expanding at a steady rate the distance of the galaxy calculated by its supernova and its redshift would tally. However, if the Universe were slowing down, as was expected at the time, we know that we would find that the supernovae would be nearer to us and would therefore, be brighter than predicted by the Big Bang theory. Much to most everyone's surprise, it was found that the supernovae were fainter, and therefore further away from us than expected. *Rather than slowing down the expansion of Universe was accelerating*. Something was pushing the Universe apart despite the pull exerted by the gravity of the matter within it.

Today, scientists know how fast the Universe is expanding and how much extra energy is pushing the Universe apart. But they do not know what this extra energy is or where it comes from. They have given it a name: *dark energy*. Dark because we do not know or understand it. You will remember dark matter from Chapters 2 and 3, which is matter we do not understand, but which we know exists because of the way the galaxies rotate about their centre. When the scientists consider all the matter and energy in the Universe together in the calculations, they find that normal matter forms only 5% of this, while the dark matter forms about 27% and the mysterious dark energy makes up the remaining whopping 68% of all the matter and energy in the Universe.

We have said that WMAP has shown that we live in a near flat Universe, one that will keep expanding, ever more slowly, forever. But will the Universe ever die, and what will happen when it does? A lot depends on what dark energy does. It is pushing the Universe into expansion. Will its strength ebb, or will it grow stronger?

We do not know. There are scenarios such as the *Big Rip*, where the acceleration increases to the point where all matter is ripped apart, and others that we have already considered where the acceleration reverses and we head to the *Big Crunch*, or the *Big Bounce*. But the view we will now review is that the ultimate destiny of the Universe is the *Big Freeze*: to expand, forever, perhaps at a changing rate, but never ever stopping.

Chapter 7. Farewell to the Universe…

How will it all end?

Infinity is forever. Forever is a long time. There is time enough for all the stars to run out of fuel, then explode or shed their nebulae and become white dwarfs which slowly cool down into black embers. Time enough for any surviving planets and stars to spiral towards each other and towards the centre of their galaxies. Time enough for them to fall into the back hole at the centre of their galaxy. Time enough for any other matter remaining to collide and break up into their constituent particles and for these particles to 'evaporate' into photons of energy. Towards the end, we will likely simply be left with black holes in a sea of energetic photons. But this Universe has one more step to go.

Stephen Hawking, of whom we have spoken before, showed that black holes are not truly black. They emit a tiny amount of radiation due to a quantum mechanics principle that in a vacuum, particles can pop into existence spontaneously. These particles always appear in pairs with one always an antiparticle twin of the other. Almost always they collide in a smidgeon of time after popping into existence and disappear.

Occasionally, though one such particle at the edge of a black hole is captured by the black hole's gravity, while the other escapes. This escaping radiation is called *Hawking radiation*. The energy involved is extremely small and the particle event is infrequent. But time is the one thing we have plenty of when we talk of infinity and, slowly but surely, over time the black holes will lose mass and evaporate. It is estimated that a black hole that has swallowed a whole galaxy will take some 10^{27} years to evaporate. Remember that is 1 with 27 zeros. A billion is 10^9 so the 10^{27} years can be written as 1 billion, billion, billion years. That is a big number, particularly when you consider that the Universe itself is less than 14 billion years old now.

[Note: Recent experiments appear to support the existence of Hawking radiation].

Thus, the Universe will keep expanding and becoming colder and colder heading towards absolute zero (0 K), the lowest temperature ever possible when all physical movement ceases. Slowly over time even the largest black holes will evaporate away, and we will be left with energy in the form of photons whizzing about at the speed of light. However, due to the infinite expansion of the Universe, the wavelength of the photons (and therefore, their redshift) will head to infinity and their frequency to zero. We know that the energy of photons depends on their frequency, so their energy will also head to zero.

It looks like the end will be slow, dark, empty, and cold. But the conclusion must be that we do not really know for certain what will happen at the end. We can conjecture. But for now, this is another one of the great unsolved mysteries of science.

But is this the final word?

We said at the beginning of our story that it all started with the Big Bang. However, scientists are very curious folk as you would have gathered by now. They have considered and put forward other hypotheses about what may have happened. Some of these hypotheses agree with the Big Bang and provide explanations as to how it may have occurred, others suggest possibilities other than the Big Bang. Some discuss that our Universe may not be the only one in existence, indeed that there may be an infinite number, universes without end.

[You may skip the following sections and return to them later]

Let me introduce you (very, very briefly) to one of the important ones of these theories. But even before we talk about the theory, we need to understand the concept of *dimensions*.

In the universe we live in, we need three measurements to clearly define an object: how wide, how long and how high it is. Similarly, when we want to describe the position of a place from where we are, we do this by measuring how far North or South it is from us, how far East or West, and finally, how far above or below us it is [Fig. 7.4]. The directions of these measurements are called dimensions. We live in a 3-dimensional space.

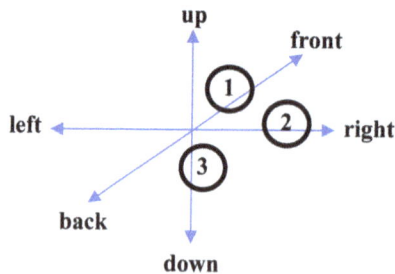

Fig. 7.4 We live in a 3-dimensional space.

Let us imagine what it would be like if we lived in a one, or a 2-dimensional space.

A one-dimensional space would only have one dimension, say, length. There would be no width or depth. An example of a one-dimensional object is a line on a sheet of paper. Not any line, but one that has no thickness at all. In the real world, if you drew a line on paper and looked at it through a magnifying glass you would see that it has width, because the point of the pen or lead in the pencil you used to draw the line has width. But imagine you had an exceptionally fine pen or pencil and managed to draw such a one-dimensional line. Anything that lived in such a universe itself would also have just length, and could only move along the line, sliding forward or backward. If you saw an ant walking on a piece of string from a distance, you may think it is living in a one-dimensional world. But if you looked at it through a magnifying glass, you would see that the thread is like a pipe and the ant has a body which also has height, length, and thickness.

Chapter 7. Farewell to the Universe…

However, a 1-dimensional object can exist in a 2 (or more) dimensional space. A string could wriggle around and move from place to place, but it would still be a string. A two-dimensional space would be like a very thin sheet of paper with just length and width. Persons that existed in such a universe would have to be flat, like a drawing. As they moved, they would again slide about but could do so not just forwards and backwards, but also sideways on the sheet of paper which is their world. They would have no concept of anything above or below the flat sheet [Fig. 7.5]. If you were to push your pencil through the paper, the two-dimensional person would find that a hole had suddenly appeared in their world, and they could only get across this hole by sliding around it.

Fig. 7.5 Flatland: A 2-dimensional world.

Let us imagine we live in Flatland [Fig. 7.5], which has only two dimensions of space. There is only forward & backward, and right & left. You could never get behind the mountain, behind would not exist, but you could walk over it as in a drawing.

As a 2-dimensional object you could happily live in a 3-dimensional space. A drawing can be rolled up in a tube, for example.

In our three-dimensional space, we also have the advantage of up & down height. Now, we think, that is a proper world. Can you imagine one where there are more than three dimensions of space? It is difficult, is not it? But that does not mean that they cannot exist. The person in the flat world would not be able to understand a three-dimensional one. But we know one exists, ours.

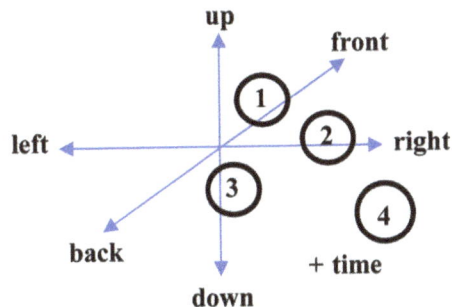

Fig. 7.6 We live in a 4-dimensional world.
Three dimensions of space + one of time.

In our three-dimensional world we can find any object by knowing three measurements. However, if you want to meet someone at a particular location in two hours, we also need to know the time. Thus, time is also a dimension. We therefore actually live in a 4-dimensional universe, three dimensions of space and one dimension of time. Fig. 7.6.

If we had access to a 4-dimensional space, we could live in it, just like the 1- and 2-dimensional objects can happily exist in higher dimensional spaces. But we could not participate in it. The man in the drawing cannot jump off his sheet of paper Universe.

As it happens, the 4th dimension that we have access to is time. This is the universe we occupy.

String theory

We have seen that in our universe all matter is made up tiny particles: electrons, protons, etc. Each of these particles is itself made of even smaller particles, till we come to fundamental particles known as quarks. But what are the quark particles themselves made of? The *string theory* says that ultimately every fundamental particle is made of one-dimensional strings. These are the tiniest things imaginable, and they have a major trick up their sleeve. They can form loops or be open-ended and they can vibrate in many ways, not unlike a violin string.

Each type of vibration of a violin string produces a musical note of a specific frequency. Each vibration of the cosmic string produces the fundamental particles of matter, energy, or force, resulting in different fundamental particles such as quarks, photons, or gravitons (a hypothetical particle of zero mass and zero charge that gives rise to gravity; it is analogous to the photon, the carrier of electromagnetic force). In this theory, the universe is a veritable cosmic symphony!

There is another point of interest about the world of strings: it has 10-dimensions, 3 of normal space, 1 of time and 6 others which are curled up, too small to be observed.

Let us spend a few minutes considering the implication of what I have just said. We live in a 10-dimensional world but are simply unaware of 6 of these dimensions. I do not know about you, but I have great difficulty in imaging even one extra dimension, leave alone six. The fact is that we are physically, mentally, and culturally not built to have a visualisation of these extra dimensions. We all, including scientists, must put our faith in two key things: *mathematics*, which we understand to be an exact science, the same over all the Universe, and *logic*. Thus, we bravely go where these two take us, and work things out even though they may be beyond our ken.

So, still what? You may well ask. Why are strings of interest to us here?

M-theory

The reason strings are of interest to us is that they are an attempt to develop a *theory of everything* (ToE), to describe all aspects of our universe in one theory and to bring together the theories of *general relativity* and *quantum mechanics*.

However, a problem developed with the string theory. It evolved into five separate versions. There inevitably came attempts to consolidate these five. Out of these arose the *M-theory*, which needs one additional dimension!

According to M-theory the universe exists on a 4-dimensional (3 dimensions of space, 1 of time) membrane (this gives the M in the title, or possibly the M refers to 'magic' or 'mystery'!). The theory says that the membrane (or 'brane' for short) is made up of 11 dimensions, 10 which we have seen above: 3 of space, 1 of time, 6 more that are tightly curled up, as in the string theory, and finally, 1 in which the brane 'floats'.

There are many models that have been proposed. In one model, two branes 'bump' into each other. If one of the branes is cold and empty, a huge release of energy occurs in it. This causes a rapid expansion in the empty brane, which is equivalent to the big bang and inflation. After this, the story unfolds as we have discussed earlier with the formation of matter, stars, and galaxies. Eventually, the universe expands, becomes cold and empty - ready for another big bang?

Other interesting proposals in the M-theory include a massless particle called the 'graviton', the quanta of gravitational field we met earlier, and the idea that dark energy results from the energy that controls the separation of the branes. Certainly, the graviton may be the subject of a search when CERN's proposed Future Circular Collider (FCC) gets built.

How do we know that the M-theory is correct? How can we ever test it? Surprisingly, a way has been suggested of checking it out. One idea is that a collision of the brane containing our universe with another, would have left a specific mark on the CMB, the Cosmic Microwave Background radiation which is found wherever we look in our universe. The CMB we know is the light from the time that the universe became transparent some 380,000 years after the Big Bang, enabling light finally to move freely across the universe. The CMB is being very intensively studied; if there are any anomalies they will be eagerly sought and analysed. Who knows what we will find?

Other theories suggest that the period of inflation which occurred immediately after the Big Bang gave rise not just to our Universe but to an infinite number of others ('multiverses'). Each of these island universes would have their own laws of physics, some of which enable stars and galaxies to form and others not. We just happen to be living in a Universe where this was possible, and one in which life could and did evolve. That is why we can ask these unending questions.

Still other theories have the Universe as a hologram, a multi-dimensional projection, parallel Universes, infinitely repeating Universes with infinite numbers of you! These are all in the realms of imagination but have roots in science and mathematics.

Will one of these win through? Our search continues.

[End of optional sections]

Einstein, spacetime and the shape of the Universe

We have seen that the answer to what is the future of the Universe is being worked on by many scientists and some exotic theories abound. One of the key contributors has been, of course, Einstein. We talk more about Einstein, his theories and thought experiments in Appendix A.

The conclusions of Einstein's work were presented in the form of "field equations" of general relativity. These equations were complex. But an exact solution was produced by four scientists: Alexander Friedmann (1888-1925) a Russian cosmologist; Georges Lemaitre (1894-1966) a Belgian Jesuit priest, astronomer, and physicist; Howard Robertson (1903-1961) an American mathematician and physicist and Arthur Walker (1909-2001) an English mathematician. The solution is called the Friedmann-Lemaire-Robertson-Walker (FLRW) metric. It is also sometimes called the Robertson-Walker metric and other names. It was based on the individual original works by these four scientists, together with Einstein's equations.

The solution is needed to answer the question whether the Universe is destined to expand, collapse, or stay steady. The prediction of the future of the Universe is based on the values of some critical factors and constants given in Einstein's results and the solutions of the equations.

The considered view at the time was that the Universe's expansion will slow down and reverse. Einstein could not believe that the Universe would be changeable. So, he introduced a factor into his equations that he called the *cosmological constant*, an acceleration that would kick in to counter the contraction. He later called this his 'greatest blunder'. But, when it was found that the Universe was expanding faster than anticipated, he realised that the factor was in fact correct, but that its exact value was not clear. The cosmological constant value depends on dark energy.

Another key factor we met above is the *density parameter*, Ω, which is the value of the average matter density in the Universe. We saw that the value of the density parameter determined the geometry or shape of the Universe. $\Omega = 1$ gives us the flat Universe, $\Omega < 1$ the open Universe and $\Omega > 1$ the closed Universe.

Chapter 7. Farewell to the Universe…

The fate of the Universe depends on the tussle between Ω and dark energy. A likely scenario is one where the Universe expands without end, but ever more slowly. But the fact is that we do not know. We need to understand what dark energy is, plus figure out how it will behave in the future. Our limited knowledge notwithstanding, the work of Einstein and the other great scientists has given us some tools to work with in trying to figure out what will be the eventual fate of our Universe.

There are a lot of scientists busy looking for the answers.

Watch this space. Or should that be spacetime.

Appendices

A. **Riding on a light beam...** the relative Universe
B. **The strange world Alice would see...** if she sat on an atom
C. **Three of my favourite scientists...** Albert Einstein, Paul Dirac, and Richard Feynman
    And one that history forgot... Rosalind Franklin

To make the subjects easier to follow, I have separated out the discussion of Einstein's theories of relativity and the theory of Quantum Mechanics into the two appendices that follow, rather than leave them in the main body of the book. You may wish to skip one or both altogether on the first reading and come back to them later. However, if you do tackle them, you should find them fascinating,

I have also included a third appendix which talks briefly about the lives of three of my favourite scientists, Einstein, Dirac, and Feynman, who had a profound impact on theoretical physics and Quantum Mechanics, and me. I hope this brief introduction will lead you on to find out more about them and their subjects.

There is one other scientist I want to bring to your attention. Her name was Rosalind Franklin. She deserved recognition for her work on X-ray photography that was essential for the success of the DNA-structure project, but sadly and, in my opinion, very unfairly, her contribution was ignored at the time.

However, justice is finally, but slowly, being done. She is belatedly receiving the accolades due to her, though long after she has died. Hopefully, the modern and future generations of women scientists will accrue the benefit from what she achieved in her brief life.

Appendix A

Riding on a light beam...

the relative Universe

Einstein was a remarkable man. He realised that while Newton's theories explained perfectly adequately how the laws of motion worked for every-day needs, they failed when faced with exceedingly high gravity and speeds near to that of light. He proposed a whole new way of looking at the Universe and developed theories explaining what happens when objects experience these extreme conditions. He showed that nothing can travel faster than light. He explained that mass and energy were different forms of the same thing and developed the famous $E = mc^2$ equation that we have already come across. He demonstrated that gravity bends light and predicted that black holes can exist. Einstein was indeed a remarkable man.

Einstein's theories are an essential part of our story since they explain how the Universe really works. His predictions have been extensively tested and not once have they been found to be lacking. His work has had a major impact on much of our modern scientific thought. His influence further extends to other major theories that followed, including the astonishing theory of quantum mechanics that we will discuss in the next chapter.

This chapter is about some of the things that we have learnt from Einstein's work. It is obviously very, very simplified. But I hope there is enough here to give you an insight into his amazing findings.

I will also tell you about some experiments that were performed which test out the predictions of these theories. To-date, there has not been any test that has come up with results that contradict Einstein's conclusions. However, his theories can be said to be incomplete. They do not cover the exceedingly small, as does Quantum Mechanics which we cover in Appendix B. Einstein tried, to the end of his life, to reconcile his theories with Quantum Mechanics, to develop a Theory of Everything (ToE), but without success. However, this effort continues.

I have given a thumbnail sketch of Einstein's life in Appendix C, together with the lives of two other of my favourite scientists, Paul Dirac, and Richard Feynman. In this chapter, I want to briefly discuss some of the thought experiments that help explain the amazing conclusions of his two theories: *theory of special relativity* and *theory of general relativity*.

I hope by the end of the chapter you will see why trying to understand the Universe without discussing relativity is like wandering in a dark night, without a torch. You see shapes and can determine roughly how large or small they are, but all colour, all sharpness and detail will be missing.

Some thought experiments

Let us do some thought experiments of our own in line with Einstein's approach. We will start with the example we have already quoted.

What happens if you ride a beam of light?

What would happen if you could ride a light beam? This was a problem Einstein pondered when he was only 16. He knew that light was electromagnetic waves travelling at 300,000 km per second. But he also knew, from Maxwell's work, that these waves always moved at the speed of light and could not stand still. He was in a quandary. For him to ride a light beam he would need to catch it, in other words to run alongside it fast enough so that it was going at the same speed next to him and he could hop on. If he managed to do this, he would see a wave that was stationary. However, this would violate Maxwell's principles that the wave always moved at the speed of light and could not stand still.

This led him to question the whole notion of relativity as it was then understood. He contemplated it for the next 10 years till he resolved it with his own *theory of special relativity*, when he was in Berne, working as a patent clerk.

However, let us follow Einstein's approach and imagine that you *could* catch a light beam and jump on. Better still, imagine that you became part of the beam, a photon if you will, massless and able to travel along with the other photons. What would you 'see'?

Let us say that your beam hits a clock in a tower just as it strikes twelve. The beam reflects off the top of the hour hand and away you go, travelling at a phenomenal speed, faster than anything else could possibly travel. Your beam of light meets nothing and nobody as it shoots beyond the Earth and out into space. It travels on and on with you into empty space at the ultimate speed carrying the information that the clock has just struck twelve. The information about one second past twelve, would be carried on the beam of light which could never catch up with you.

However long you travel and how far the beam goes, your time does not change. If you had a watch, it would say twelve o'clock. It is always twelve o'clock, always NOW, for this light beam and for you in the beam. Everything is frozen at that instant. Time has stopped. But on Earth, for every 300,000 km you travel, one second passes. In just over a second you have shot past the moon, in 8 minutes you are beyond the Sun. You travel on and on. Years pass on Earth, then centuries. For you it is still twelve o'clock on the day you reflected off the tower clock.

Finally, after a thousand Earth years you reach a solar system, with its own sun with an Earth-like planet in orbit. You shoot towards this planet and head towards a land mass on which there is a house with a young alien female looking towards the Earth through, what is for us, an amazingly advanced telescope, powerful enough to see the clock tower from where you came.

Appendix A. Riding on a light beam…

Your light beam goes through the telescope into the girl's eye. Her brain gets the signal from the eye, and the girl exclaims, "Oh look. It is twelve o'clock on the clock on Earth". For you, no time appears to have passed, but on Earth 1000 years have gone by. You have travelled for 1000 years at the speed of light, to reach a distance we call 1000 light-years. Interestingly, the information about it being twelve o'clock has also travelled at the same speed, showing that even information cannot travel faster than this ultimate speed.

Now think about what we have just said. For a light beam, it is always NOW. Time does not pass. So, when you became a photon and joined the beam your speed jumped from 0 to light speed, your personal time went from your normal time of 1 second passing every 1 second to 0 seconds passing every 1 second. You can conjecture that if your speed had gone from 0 to half the speed of light, perhaps your time would have slowed down not to 0, but maybe to somewhere between 0 and 1. And you would be right. This slowdown of time with acceleration is called *time dilation*. It is one of the fundamental principles that Einstein's theory predicted. The faster you go, the slower your time becomes *as measured by someone who is not moving*. But the slowdown is not even: if you travel at half (50%) the speed of light, your time (as measured by a non-moving observer) slows down by some 13%, at three-quarters (75%) the speed of light, the time slows down by 34%, and at 90% of the speed of light, the time slows down by just over 56%. Of course, at 100% the speed of light the time slows down by 100% to 0.

In fact, it is not the speed that causes time to slow down, but the acceleration. The time slows down for our intrepid light-traveller because his speed jumps from 0 to near light speed; once he is travelling at a steady speed, no such change occurs. He retains his new time till he slows down (decelerates) back to normal speed and normal time.

Let us now look at a couple of other thought experiment that show time dilation.

What happens if you bounce a light beam in a train?

Suppose that you have a friend who is riding in a train carriage [Fig. A.1]. The train passes a station at a steady speed without stopping. It just so happens that you are on the station platform and able to see your friend in her carriage as it goes by. You see that she is bouncing a ball on the carriage floor. To her the ball is going straight up and down as it bounces on the floor of the carriage and back into her hand. But what will you see?

The train moves forward past the station platform. By the time the ball leaves her hand and strikes the floor, the train has moved forward, say, by 5m. Again, as the ball bounces back into her hand, the train moves forward another 5m. To you the ball looks as if it is moving down from her hand to the floor and back up again some 10m from where it left her hand. You can see this in Fig. A.1. The important point is that you see the ball travelling a distance that is more than the straight up and down distance between each bounce.

So far, so good. But what does this have to do with light?

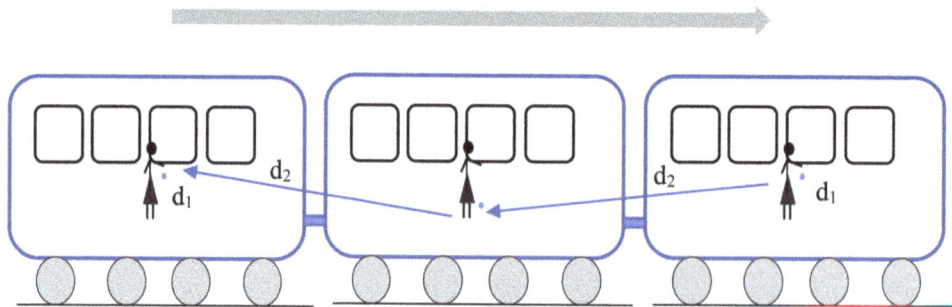

The man outside the train sees the ball travel $d_2 + d_2 = 2d_2$ taking the same time as the girl on the train sees her ball travel $2d_1$ (d_1 down + d_1 up).

Fig. A.1 A bouncing ball in a train travelling at speed. The ball appears to travel further to an observer outside the train than to one in the train.

Well, now imagine that instead of a ball, your friend is bouncing a beam of light between two mirrors, one in her hand and the other on the carriage floor. Imagine also that you can see the light beam as it bounces up and down and can measure the path the beam takes. The light will behave just like the ball and seem to you to be travelling a larger distance than the straight up and down distance between the two mirrors. But to your friend the light beam is still travelling the down-and-up distance. Now consider that this arrangement is a simple clock where each reflection of the light beam between the two mirrors is a tick of the clock. You and your friend decide to measure the time of this tick, with her aboard the train and you on the platform.

In the time that the light beam has travelled up and down as measured by your friend, it would appear to have travelled a longer distance as measured by you stationary, on the platform, just as the ball did earlier. Since the speed of light is constant the time taken must therefore be different in the two cases. When you and your friend meet up, you compare the times you recorded and find that your recorded time is longer than your friend's. If you are measuring what you think is the right time since you are stationary, your friend's watch must be going faster since you are measuring a longer period. However, as far as she is concerned, her watch is correct, so it must be your watch that is going slower. Whose time is right?

Appendix A. Riding on a light beam…

Well, both are, from individual points of view. However, time dilation depends on who has been moving relative to the clock. The clock in this case was the mirror and light arrangement on the train, travelling with your friend. Therefore, you are the one that has been moving *relative to the clock* and her time is right. *Clocks which are at rest relative to an observer run faster than those that are moving relative to the observer.*

Note that we have been talking of a train moving at a steady speed with no acceleration. When you take acceleration into account, even stranger things happen.

Let us illustrate this with the famous twins' paradox.

The twins' paradox – what happens when one twin goes for a ride and the other stays on the farm?

Suppose we have a pair of twins who are 25 years old. One is a farmer, the other an astronaut. An opportunity arises to test a newly developed space drive that can reach almost the speed of light. The astronaut twin volunteers for the mission and is selected. After a tearful farewell to her brother, she is on her way.

The spaceship accelerates reaching 95% the speed of light. After two and a half years of travel, the twin decelerates the spaceship, turns it around and accelerates away again to almost light speed heading back to Earth. On her return, she finds a quite different world. Things have dramatically changed. She cannot recognise the place. She looks at a newspaper and discovers that over 16 years have passed since she left. Yet, as far as she is concerned, she has travelled for only five years, and is only 30 years old. She appears to have travelled into the future.

She tracks down her twin brother. He is still at his farm. But she finds a 41-year-old man, with a family of teenage children. What happened? Time dilation happened.

The spaceship was a resounding success. It reached almost the speed of light. Time slowed down for the twin on board by almost 69%. Each of her years was equal to more than 3 years on Earth, as measured by clocks (and newspapers) on Earth. If she had travelled even faster, or longer, the slowdown would have been even more dramatic.

Is the twin paradox scenario likely?

As far as we can tell, such a high speed is impossible for humans. Why? Because of an associated relationship that was also shown by Einstein. As speed approaches the speed of light, time will indeed slow down towards zero, but the mass will increase towards infinity.

In fact, this is a reason why nothing can travel faster than light. As an object gets nearer and nearer to the speed of light, it gets more and more massive. Remember Einstein's famous equation: mass and energy are different aspects of the same thing and can be changed, one into the other. The fast-travelling object gains its additional mass due to the enormous energy it has obtained by travelling at these high speeds. You can say the object gains weight, though weight is the consequence of a force by which something, such as the Earth, pulls the object towards itself. Therefore, weight can only be relevant where gravity comes into play. The gain in mass by the object means that it requires ever increasing amounts of energy if it wants to go even faster. Ultimately, if it approaches the speed of light, its mass heads towards infinity and it will require an infinite amount of energy to gain that extra 5% speed beyond the spaceship's 95%. This is, of course, impossible, so the speed of light cannot be reached by anything that has mass.

A different limit holds for particles which have no mass, such as photons which are the particles carrying light. The only *speed* that such massless particles can travel at is the speed of light. That is why light travels at the speed that it does, no faster and no slower. It cannot do anything else.

Length contraction

We have seen that as objects travel near to the speed of light, time slows down for them, and they gain mass. There is another strange effect that the object undergoes. Their length shrinks! Or at least it does when it is measured by an outside observer. A spaceship heading for a star at 95% of the speed of light will shrink by some 69%. (Note that this is the same ratio that earlier we found applies for time dilation.) It follows that if the spaceship were to reach the speed of light its time will go to zero, its mass will become infinite, and its size will become zero. Warp drive appears to be out, I am afraid.

But here is a fascinating question: what happens to the astronauts on the spaceship? They do not feel time is slowing down for them at all. As far as they are concerned, a second is still passing every second, on *their* clocks. Their ship is the same size it always was *by their measurements*. What is going on? How come that the astronauts travelling at their speed think they reach their destination and get back in less time than the people left behind think they have taken. Well, for the accelerated astronauts, time still appears to pass at the same rate, *but the distance that they travel has shortened.*

If astronauts fly away from us at near the speed of light, we will see their clocks ticking more slowly than ours on Earth. However, to them their clocks are still ticking at the same rate, but ours are ticking faster. For them, a second still passes every second, but they are travelling a smaller distance. Therefore, in the twin paradox we saw earlier, the travelling twin comes back finding less time has passed for her because she has travelled a shorter distance, while the twin on Earth knows otherwise. This may seem very strange, even unbelievable. But it is demonstrable, as we shall see later in the chapter when we talk about muons.

Appendix A. Riding on a light beam…

Do we always agree on when things happen?

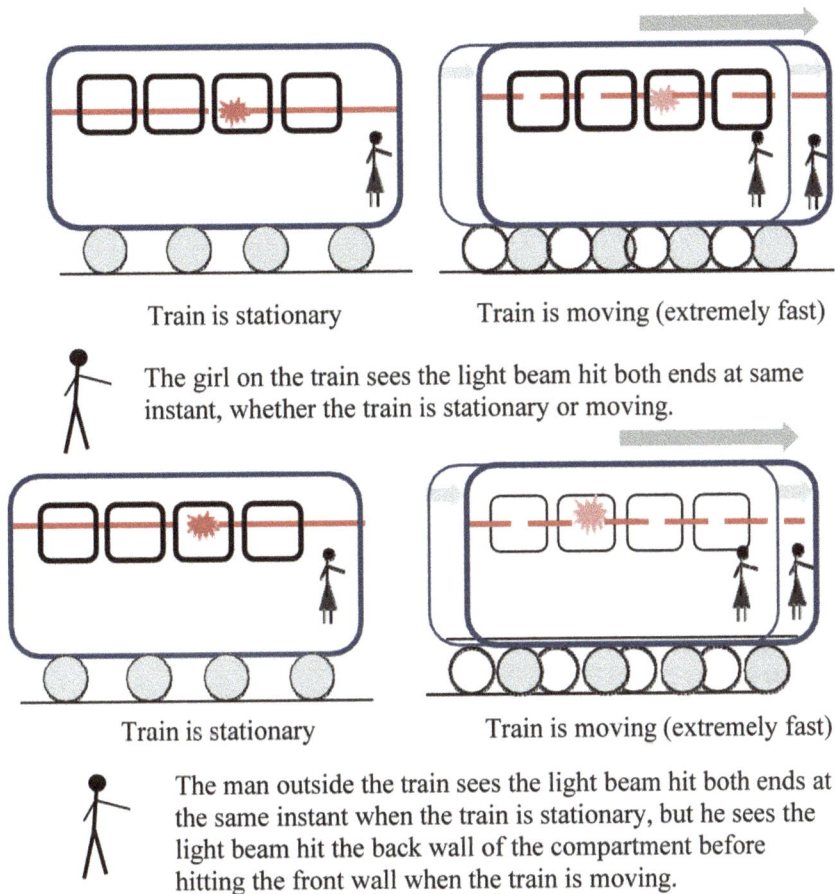

Train is stationary   Train is moving (extremely fast)

The girl on the train sees the light beam hit both ends at same instant, whether the train is stationary or moving.

Train is stationary   Train is moving (extremely fast)

The man outside the train sees the light beam hit both ends at the same instant when the train is stationary, but he sees the light beam hit the back wall of the compartment before hitting the front wall when the train is moving.

Fig. A.2 Observers can disagree when simultaneous events occur.

Let us perform another thought experiment relating to Einstein's theories. This time a train is standing at the station, and there is only one carriage from which all furniture has been removed, so there are no seats, just an empty shell. In the middle of this carriage there is a device that can send two flashes of light, one towards the front of the carriage and one towards the back. Let us assume also that we have amazing instruments that can react fast enough to see any differences between the times that the light hits the front and back of the carriage.

There is someone on board who can set off these flashes. You are standing on the platform as usual, observing [Fig. A.2].

With the train stationary, the person on board sets off a flash. As we would expect, the light beams hit the front and back of the carriage at the same instant. We all agree on this.

But now consider this. The person on board sets off another flash when the train is moving. What will each of you see now? Let us think about it.

The person on board will see the light beam hit the front and back of the compartment at the same time. However, for you as the flash goes off, the back of the train is moving forward towards the flash while the front of the train is moving away from the flash. Thus, you (with your precision instruments) will see the flash that is going towards the rear of the train strike the back wall first, before the flash going forward has hit the front wall.

Therefore, you and the person on board will not agree whether the light beams hit the two walls simultaneously or not.

What happens when we ride up in a lift?

Let us consider another of Einstein's thought experiments, this time related to gravity.

Suppose you are standing in a lift in a very tall building. You press the button for the top floor. The lift starts moving upwards, slowly at first, then faster and faster. What do you experience? As the lift moves up, you are pressed into the floor more and will feel heavier. This feeling increases as the lift speeds up, then stops as the lift climbs with a steady speed. Near the end of its journey, the lift slows down, and the floor falls away, and you feel lighter till you stop.

Suppose now you are still in the lift but this time standing on electronic scales that measure your weight. Let us say that the scale reads 70 kg with the lift stopped. If you looked at it as the lift rose, speeded up, stayed steady, and then slowed down and stopped, the scales reading would increase to perhaps 75 kg, reduce, and stay steady at 70 kg, then drop to, say, 65 kg, before settling back again at 70 kg. You gained weight when the lift was speeding up (accelerating), stayed steady at your normal weight when it was climbing at the same steady rate (no acceleration), then you lost weight while the lift slowed down (decelerated) before becoming steady again when the lift stopped.

I am sure that we have all been through other such sensations, for example, on a roller coaster ride. The difference with Einstein was that he reasoned that one could argue that gravity and acceleration were the same thing. He reasoned that if a spaceship, moving in space away from the influence of Earth's gravity, was accelerating continuously at the rate with which the Earth attracts all objects, you would not be able to tell, if you did not already know, whether you were travelling in an accelerating spaceship in outer space or were stationary on a large body such as the Earth. When this spaceship stopped accelerating, you would become weightless and float.

Appendix A. Riding on a light beam…

If you are in a satellite going around the Earth, the Earth is still attracting you, and you are still falling down to Earth (as is the satellite), but, because of your forward motion, you are falling at an angle that mirrors the Earth's curvature. That is why astronauts are weightless in the International Space Station.

Einstein said that there is no difference between the cases where a body is being accelerated at the rate of Earth's gravity, compared with the case where a body is acting solely under the influence of Earth's gravity. He argued that any gravity-dependent experiment you perform in the two situations will come up with an identical result in each case.

An interesting point occurs if gravity and acceleration can be thought of as the same thing. Just as acceleration can slow time (time dilation), gravity should be able to do the same. In fact, this is exactly what is found. Time runs more slowly in locations where the gravity is high. In other words, a clock that is at the bottom of a mine shaft (which being nearer the centre of the Earth has a higher gravity) will run slower than a clock on the surface of the Earth, which in turn will run slower than a clock at the top of Mount Everest. Experiments performed with exceptionally accurate atomic clocks have shown this to be so.

We can measure the acceleration due to gravity here on Earth. In science, this acceleration is given the symbol g. On Earth the value of g is approximately 9.8 m per second per second (also written as 9.8 m/s^2). In other words, the Earth pulls *all* objects towards its centre with an acceleration that increases the falling speed by 9.8 m per second every second. So, the objects start to fall with a speed of 9.8 m/s. After one second the speed has increased to 19.6 m/s (that is, 9.8 x 2). After another second, the speed has become 29.4 m/s (9.8 x 3), and so on. Eventually, air resistance stops the increase in speed.

This means that all objects falling due to gravity fall at the same rate. As we have already noted, this fact was demonstrated by the famous Italian scientist Galileo Galilei (1564-1642) in Pisa around 1589 when he dropped two similar balls, one much heavier than the other, off the leaning tower of Pisa. They hit the ground at the same time.

The question is often asked: does that mean that a feather will fall at the same speed as a hammer? The answer is yes. It does not happen when you drop the two here on Earth, because the air in our atmosphere affects the lighter feather more than the hammer. However, an experiment done on the Moon by Commander David Scott on the Apollo 15 mission in 1971 showed this to be the case. You can see the video on the internet of the experiment being performed.

What Einstein's theory showed was that if you are in a rocket (far away from the Earth's gravitational influence) accelerating at 9.8 m/s^2, not only would you weigh the same as you do on Earth, but all things on the rocket would behave as on Earth - if you dropped a ball it would fall to the rocket's floor exactly as if you were dropping it in your room at home [Fig. A.3].

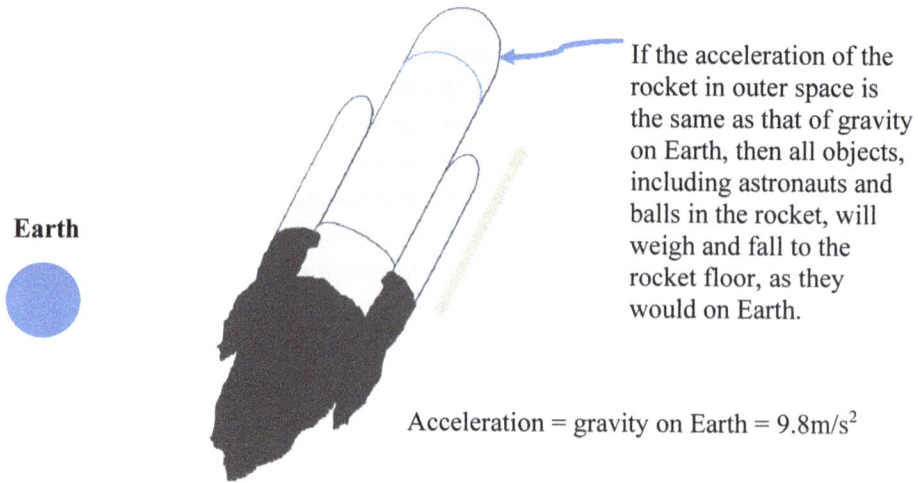

Earth

If the acceleration of the rocket in outer space is the same as that of gravity on Earth, then all objects, including astronauts and balls in the rocket, will weigh and fall to the rocket floor, as they would on Earth.

Acceleration = gravity on Earth = 9.8m/s^2

Fig. A.3 Acceleration and gravity 'are the same'.

The Moon's gravity (g) is less than the Earth's since the Earth is bigger and more massive than the Moon. In fact, the acceleration due to gravity (g) on the Moon is approximately 1.6 m/s^2, about 6 times less than that of the Earth. This means you will weigh six times less on the Moon than on the Earth and jump that much higher and further. Your rocket would only need to be accelerating at 1.6 m/s^2 for you to weigh the same as on the Moon, and for a feather and a hammer to fall as it would on the Moon (with no air in the cabin on the rocket, of course).

How does gravity affect light?

Let us again imagine that we are riding in a rocket coasting along in space at a steady speed. As there is no acceleration, we will feel no force on us and will be weightless. Now imagine that we shine a laser beam from one side of the rocket cabin across to the opposite wall [Fig. A.4 (a)]. The laser light will shine on the wall across the cabin from the point where the beam started. We mark the point.

Next, we accelerate the rocket. As we have already discussed above, in this case we will feel a force pushing us towards the base of the rocket which is like the effect that gravity would have.

Now comes the interesting part. Imagine that we again shine the laser from the same point as before across to the other side of the cabin. During the time that the laser light takes to go across the cabin, the rocket would have accelerated forward ever so slightly. The laser beam will therefore hit the wall a tiny bit below the original strike point. We are assuming that in our imaginary experiment we have instruments accurate enough to mark and measure this minute distance

Appendix A. Riding on a light beam…

[Fig. A.4 (b)]. As we can see from the two diagrams, in the case of the accelerating rocket, the light from the laser has been bent (much exaggerated here of course). Since we can take acceleration and the force due to gravity to be the 'same', we can say that gravity bends light.

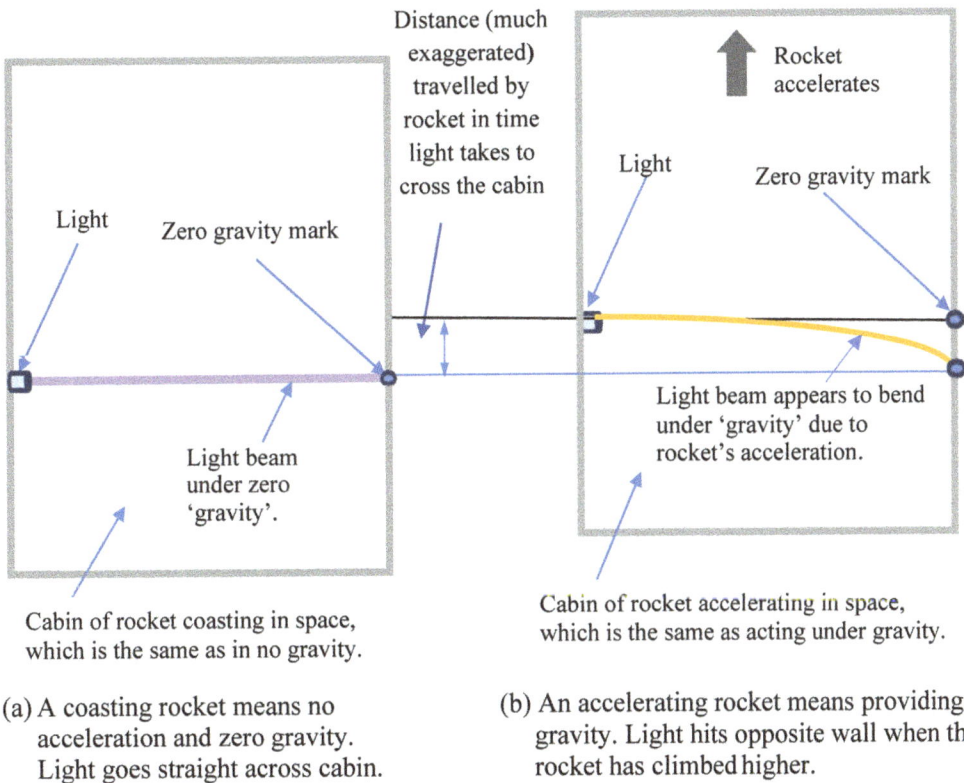

(a) A coasting rocket means no acceleration and zero gravity. Light goes straight across cabin.

(b) An accelerating rocket means providing gravity. Light hits opposite wall when the rocket has climbed higher.

Fig. A.4 A thought experiment to show that 'gravity bends light'.

Gravity also affects the frequency of light in another way. If light travels from a point of high gravity to one of lower gravity, it is redshifted. In other words, it is as if the light 'has to work' to overcome gravity that is trying to hold it back. We know that the energy in light is shown by its frequency: blue light has a higher frequency and more energy than red light which has a lower frequency. Thus, when a light beam must overcome gravity, it loses energy and becomes redder; its wavelength increases and its frequency decreases. We say that the light has been 'redshifted'.

This can occur, for example, when light is shone from a massive neutron star into space. It is as if the higher gravity was pulling on the light beam and stretching its wavelength. The opposite is also true: light is 'blueshifted' when shone from a mountain top into the valley below, its frequency (energy) increases, and its wavelength decreases. In other words, it gains energy as it goes from a low gravity to a higher gravity location [Fig. A.5].

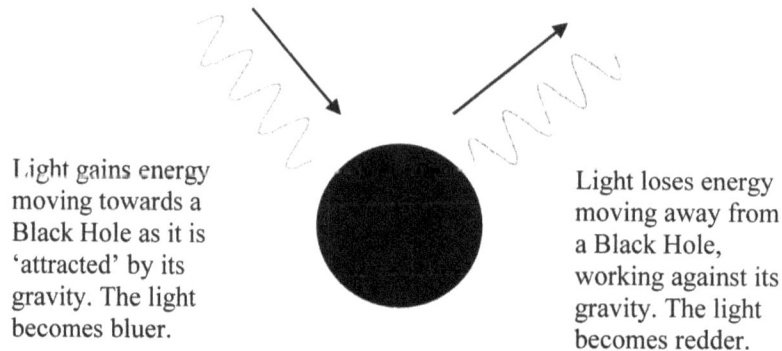

Light gains energy moving towards a Black Hole as it is 'attracted' by its gravity. The light becomes bluer.

Light loses energy moving away from a Black Hole, working against its gravity. The light becomes redder.

Fig. A.5 Gravity affects light's energy (frequency).

Near a black hole with its extreme gravity, any light that just manages to escape from a near encounter with the beast will consist of extremely long waves, and an extremely low frequency (energy). Light that is within the black hole's event horizon does not escape at all. One way of looking at it would be that it just does not have the energy to overcome the gravitational attraction.

We have seen some interesting examples of thought experiments that should help us understand some of the things that Einstein proposed in his *theory of special relativity* and *theory of general relativity*. Of course, there are many other aspects to the theories that we have not covered in this much simplified explanation. But hopefully, it has helped you understand some of the features of Einstein's theories and to appreciate the immensity of the intellect of the man.

Let us now bring together some of these key factors from Einstein's theories.

The consequences of Einstein's theories on gravity

There are several fascinating consequences resulting from Einstein's theories, some of which we have met, which are outlined briefly:

- Gravity cannot be distinguished experimentally from acceleration.
- Gravity affects time, in the same way as acceleration; the larger the acceleration or gravitational force at a location, the slower that time (and clocks) run for all at that location. This is called *time dilation*.

Appendix A. Riding on a light beam…

- Gravity affects the frequency (energy) of light. If light travels from a point of high gravity to one of lower gravity, it loses energy and is redshifted, its wavelength increases and its frequency decreases. The opposite is also true; light is blueshifted when travelling to a higher gravity location.
- Gravity can bend light.
- Nothing, but nothing, can travel faster than light.

How do we know? … Experiments and tests of Einstein's theories

Speed of light

Before Einstein, it was believed that everything needed a medium through which to travel (e.g., sound through air, fish through water, etc.). It was argued that light also needed a medium through which to move, and therefore it was concluded that space must be filled with a substance through which light travelled. This substance was called *luminiferous aether*. It was further argued that because of this, all measurements of the speed of light could only be made relative to the speed of this mysterious *aether*.

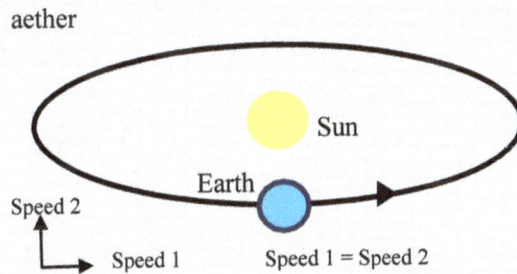

Fig. A.6 Michelson's experiment.

In 1887 two American physicists, Albert A Michelson (1852-1931) and Edward Morley (1838-1923), set out to detect the existence of aether. They reckoned that as the Earth moved in its orbit around the Sun, it would be moving through the aether [Fig. A.6], and therefore the speed of light measured on Earth in this direction would be slightly slower (by an amount equal to the Earth's speed through the aether), than when the speed is measured in a direction perpendicular (at right angle) to this direction.

Michelson and Morley used an instrument invented by Michelson called an interferometer to compare the speeds of light as measured in these two directions. The interferometer is a device [Fig. A.7] that splits a beam of light into two distinct beams which then travel along two perpendicular paths and are finally reflected by mirrors to a detector to be compared. Any difference in the two beams at the detector results in an 'interference' pattern of dark and light bands that depend on how the two waveforms line up. The differences that can arise from subtle shifts in the waves due to the reflection and refraction through the mirrors and glass and any other mechanical reason, were made the same in the two directions and cancelled out. Only differences in the speed of the two light beams due to the aether and the Earth's speed through space were measured. Much to everyone's surprise, Michelson and Morley found no shift in the pattern at all. They had shown that there was no such thing as aether, ringing the death knell for the associated theory about the luminiferous aether.

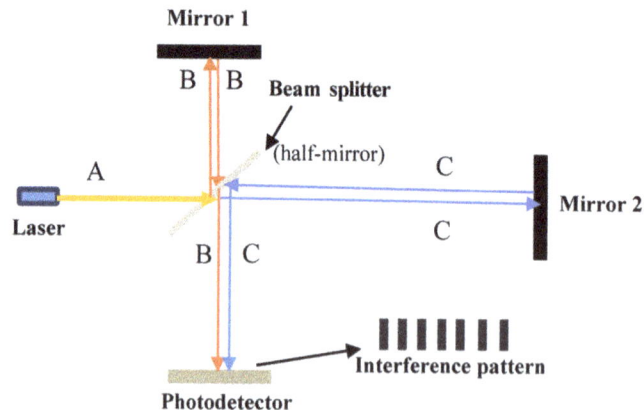

- The beam-splitter is a half-mirror which splits beam A into beams B and C.
- The mirrors reflect the beams falling on them.
- The photodetector detects the interference pattern between two beams of light due to any mismatch between the waveforms of beams B and C. The result is a highly accurate measurement of any difference between beam B and beam C due to its path to Mirror 2 and back.

Fig. A.7 The Michelson interferometer.

Michelson went on to calculate the speed of light in 1879. He was awarded the Nobel Prize for Physics in 1907.

Appendix A. Riding on a light beam…

The wobble in Mercury's orbit

All planets move in ellipses around their sun. An interesting analysis was done with the orbit of Mercury around our Sun. It was known that Mercury showed a very slight shift with time in the orientation of its orbit, but no one had been able to explain this fully. You may have seen a top spinning and noticed how it starts to wobble as it slows down, or you may have seen how the axis of a gyroscope moves as it spins. The effect is similar in Mercury's case. The effect, which is known as precession is tiny (only about $0.159°$ *per century*).

Part of the cause can be accounted for using Newton's approach to gravity. However, it needs Einstein's *theory of general relativity* to fully explain the effect.

Bending of light due to gravity

John Archibald Wheeler (1911-2008), an American physicist, once famously said: "matter tells space how to bend; space tells matter how to move".

Einstein's theory predicted that mass distorts spacetime such that anything that is travelling in a straight line near to the mass will have its path curved due to the distortion. This prediction covers the paths followed by planets, such as the Earth, in their orbit around the Sun, as well as the path of light when travelling on a path grazing a star, such as the Sun.

Fig. A.8 visualises how the Earth warps spacetime imagining spacetime as a sheet of rubber and Earth as a heavy ball resting on it. Fig. A.9 illustrates the bending of the light from a distant star by the Sun.

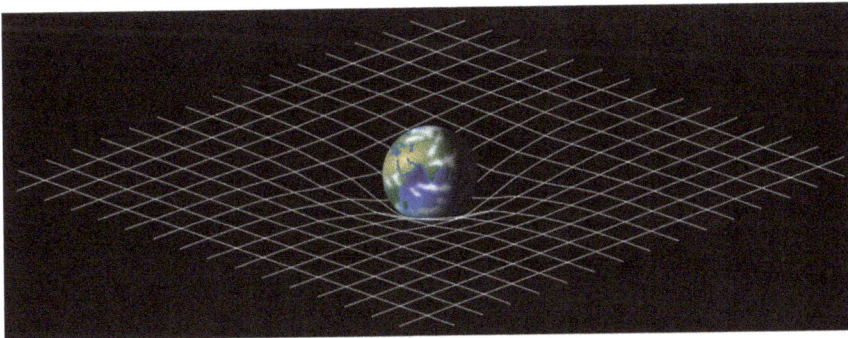

Fig. A.8 Mass distorts spacetime.

An especially important experiment, which confirmed Einstein's predictions of the curvature of spacetime by a massive object, was performed by the eminent British astronomer Sir Arthur Eddington (1882-1944) in 1919.

There was a solar eclipse that year visible on the west coast of Africa. Eddington realised that he could use that opportunity to see whether Einstein's prediction about massive objects bending light would be in line with the experiment. The argument went like this: if the position of a star shining next to the edge of the Sun was accurately measured when the Sun was covered by the Moon during the totality period, and then compared with the position of the same star in the sky when the star was in the same position relative to the Earth but at night-time when the Sun was not around, any shift seen in the star's apparent position would be due to the gravity of the Sun bending the star's light and could be calculated. This bending is explained diagrammatically in [Fig. A.9]. Eddington performed the measurement during the 1919 eclipse on the island of Principe, off the west coast of Africa.

Einstein's predicted shift in the position of the star is tiny (only about 5 ten-thousandths of a degree!) and Eddington's equipment was not exactly accurate, compared with what is available today. However, enough of the shift was detected for Einstein's prediction to be declared as confirmed. Einstein became an instant, world-wide celebrity.

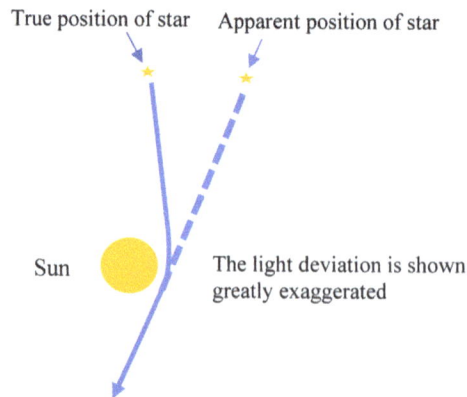

Fig. A.9 Eddington's experiment.
Gravity bends straight paths such as that of light.
This was used by Eddington to prove Einstein's assertion.

Today, the availability of radio telescopes enables experiments to measure the gravitational shift to be carried out with much greater precision (a millionth of a degree or better). These telescopes detect radio waves coming from objects in space. The discovery of the extremely bright radio emitting quasars (which we have covered earlier in the book) meant that measurements could be made of the radio signals, passing close to the Sun, during daytime, without waiting for total eclipses. These measurements have confirmed that Einstein's predictions of the effect of gravity on light are correct to better than 1%.

Appendix A. Riding on a light beam…

If you were to reproduce the rubber sheet and Earth arrangement shown in Fig. A.8, and then roll a small glass ball past the edge of the Earth, the ball will be seen to bend analogous to the bending of light by a star shown in Fig. A.9.

Another example of the bending of light by massive objects in space is called *gravitational lensing*. You would know that lenses can bend light and are used to focus light beams. They are extensively used particularly in cameras, binoculars, people's eyeglasses, and telescopes.

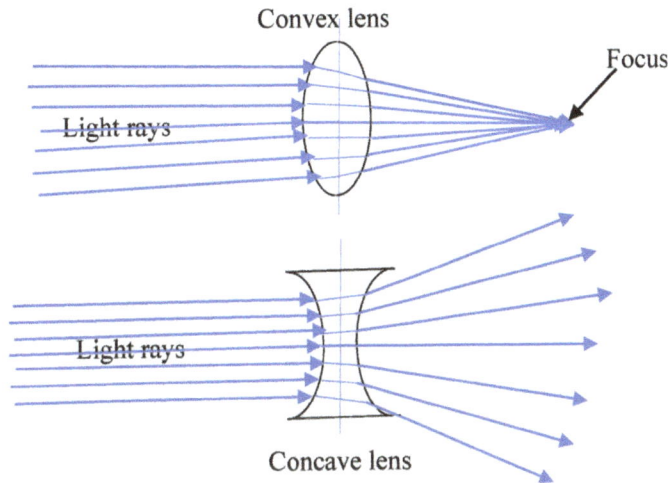

Fig. A.10 Concave and Convex lenses.

There are essentially two types of lenses, called convex and concave, depending on which way the glass of the lens curves [Fig. A.10]. The paths of the light rays through the lenses are shown in the diagram.

Imagine that instead of the convex glass lens we have a very massive object, such as a galaxy, in the middle of the 'lens'. Again, imagine that far in the distance, behind the lens, lurks a very bright object such as a quasar. In theory, we should not be able to see the quasar since its light would be blocked by the intervening galaxy.

However, according to Einstein's theory, the intervening galaxy would bend the light from the eclipsed quasar, rather as a convex lens does in Fig. A.10, and we would see light shining from the quasar on either side of the eclipsing galaxy [Fig. A.11]. If the far object and the front 'lens' were suitably aligned, a complete ring of the image of the far object would be expected to form. This is called an Einstein ring.

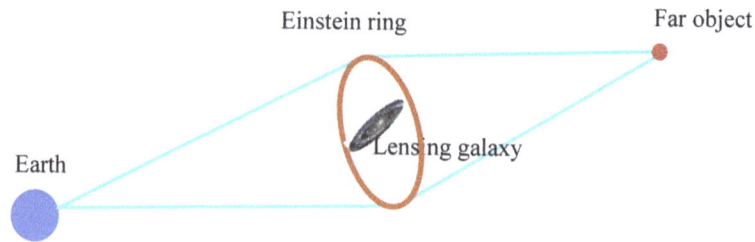

Fig. A.11 How Einstein rings form.

The gravity of the galaxy acts as a convex lens to focus the light from an object (galaxy or black hole) behind into a ring pattern around the galaxy.

Fig. A.12 How an Einstein ring looks through a telescope on Earth.

In 1979, astronomers found two quasars near each other. When they examined the spectra of these objects, they found them to be identical. They also found that these quasars brightened and dimmed together. The penny dropped and it was realised that the two spectra were of the same quasar whose light was being focused by a massive object, a dim galaxy, which lay between the quasar and the telescope. Since then, many other such examples have been found, including those of complete or almost complete Einstein rings mentioned above.

The ring in the diagram [Fig. A.12] is the image of a massive, bright object (such as a quasar) behind the focussing galaxy which is seen shining at the centre of the ring. Gravitational lensing allows us to 'see' and study objects far more distant than we would otherwise be able to.

Appendix A. Riding on a light beam…

Gravitational waves

Einstein predicted that extremely violent events, such as pulsars or black holes colliding, black holes being formed by a supernova explosion of a massive star, or even the Big Bang itself, would result in the production of 'gravitational waves' in space. Gravitational waves are ripples in spacetime, in other words, ripples that change the structure of space itself as they pass through.

Attempts have been made to detect these waves, and to prove (or disprove) Einstein's prediction. The trouble is that changes in space due to these waves would be extremely small (around 10^{-19}m, less than a billionth of the width of an atom!) and therefore exceedingly difficult to record. But that has not stopped the scientists from trying. A sophisticated and much enhanced version of the Michelson interferometer is used for this purpose.

Such an instrument has been described above [Fig. A.7], when we were considering the experiment to measure the speed of light. In this case, to detect gravitational waves, the instrument uses laser beams that are split into two beams travelling perpendicular to each other. The beams are then reflected and compared at the photodetector as in Michelson's experiment. A laser is used because of its extremely high-frequency and short-wavelength beams which provide the necessary high resolution.

As we saw earlier, the interferometer essentially compares the waveforms of the two beams which follow different paths through the device, to check whether there has been any change over beam paths. Analysing the 'interference' pattern caused by any difference in the waveforms (for example, by any changes in lengths of the arms due to the gravitational waves), indicates whether and how this pattern has been disrupted. By measuring the disruption scientists can determine the change that has occurred in the path of the beam. Since we are dealing with the waves of the laser light whose wavelength is extremely small, minute changes can be measured. Of course, the whole experiment must be very accurately and sensitively conducted to eliminate extraneous issues such as vibrations. Also, since we are measuring exceedingly tiny effects, the 'arms' of the paths of the laser beam have to be exceptionally long. All this makes for a complex arrangement.

Experiments called the *Laser Interferometer Gravitational-Wave Observatory* or LIGO have been set-up by universities in the US, including Caltech and MIT. These are based on interferometers, in principle like the Michelson instrument described earlier and illustrated in Fig. A.7, above, but have arms 4km long. They have been installed in the US states of Louisiana and Washington.

On 11 February 2016, the LIGO project published a paper confirming that analysis of signals detected by the LIGO experiment on 14 September 2015 showed they were caused by gravitational waves generated by the collision of two black holes, each about 30 times the mass of our Sun. Since then, many gravitational wave events have been detected by LIGO. The latest to the time of writing was in May 2019, when the LIGO experiment announced the detection of gravitational waves from a collision between two neutron stars or a neutron star and a black hole.

There are plans afoot, by the European Space Agency (ESA), to set up a similar experiment in space, called the *Laser Interferometer Space Antenna* (or LISA). LISA will have 3 satellites in orbit with laser arms each 2.5 million km (!) long [Fig. A.13]. In space, LISA will have the added advantage that there will be no problems due to lorries rumbling by. However, the current launch date is not till 2034.

Fig. A.13 LISA, the space-based version of LIGO.

Stop press: On 3 September 2020, newspapers announced that a collision between two huge black holes had been detected out in space. The collision occurred 7 billion years ago, when the Universe was half its current age, but the gravitational waves resulting from the event only recently reached us. They were detected on 12 April 2019 by a collaboration between three LIGO instruments, two in the USA (see above) and the third at the Virgo observatory in Pisa, Italy. The data has been under analysis since then.

We now know that one of the black holes had a mass 66 times that of the Sun, and the other was of 85 Solar masses. After the collision, one massive black hole remained with a mass equivalent to 142 solar masses. This means that the merger released around 9 solar masses of energy. Scientists believe that the larger of the two black holes was too large to have formed on its own (by a supernova) and was probably the result of an earlier merger of two other smaller black holes. It is possible that the supermassive black holes at the heart of most galaxies were formed in a similar manner.

Appendix A. Riding on a light beam…

Effect of acceleration on time and distance

We have spoken about the effect that acceleration has on time. We have also spoken about acceleration affecting the distance travelled by spaceships. Let me tell you how scientists have tested this.

High energy particles from space, called cosmic rays, are continuously hitting the Earth's atmosphere. When these particles collide with atmospheric molecules, about 15 km above the Earth, other particles are produced, including particles called *muons*. A muon is like an electron; it has the same electric charge but is over 200 times more massive.

Scientists know that a muon has a truly short lifetime (about 2 millionths of a second, or about 0.000002 seconds), before it decays into an electron and other particles. Scientists also know that normally, muons should decay into these particles well before they reach the Earth's surface and so should not be detected at ground-level at all. Yet we find them at the surface. The question is: how can that be? How is it that the muons are living long enough to reach the Earth?

It turns out that the collision with the cosmic particles accelerates the muons to within 99.98% of the speed of light. We can calculate, using Einstein's theory, that this should slow down the time for the muon (compared with our clocks), long enough for it to be able to reach the Earth. Therefore, Einstein is right. As far as the observers on Earth are concerned, time slows down for the muon as it is accelerated to near the speed of light. But here is a fascinating conundrum we discussed earlier: what happens to the muon itself? It does not feel time is slowing down for it. As far as it is concerned, after 2 millionths of a second has passed it dies. What is happening here? Well, for the muon, while time passes at the same rate, *the distance to Earth's surface shortens*. It can reach the Earth before it dies because the Earth just got closer!

We can also undertake experiments here on Earth to determine the effects of acceleration using machines such as the Large Hadron Collider (LHC), operated by CERN, the European Organisation for Nuclear Research, in Geneva, which, as we have discussed earlier, collides beams of particles that have been accelerated to near light speeds and analyses the results. The experiments using the LHC and other colliders have shown that the accelerated particles in these experiments live longer than they would be expected to, exactly in line with Einstein's predictions. Further, they have also shown that the particles gain mass (become more massive) as they are accelerated, again exactly in line with the predictions.

Other experiments that have confirmed Einstein's time dilation theories include sending highly accurate atomic clocks on jet planes (travelling at a higher acceleration than that on Earth) around the world, when the clocks have been found to run slower than clocks left behind, exactly in line with predictions. Further, clocks that have been taken up mountains and in space orbits (where the gravitational force is lower than on Earth), have also been found to run faster than Earth-bound clocks, showing that the larger the gravitational pull or acceleration, the slower the time flows.

Indeed, this difference in the rate that time passes in space compared with that on Earth must be considered in producing satellite-generated GPS signals which are used by smart phones to give us our position here on Earth.

Other Einstein concepts

Wormholes

Einstein's work also introduced the idea of *wormholes* in space [Fig. A.13]. Wormholes are intensely curved parts of spacetime that connect two points in our Universe.

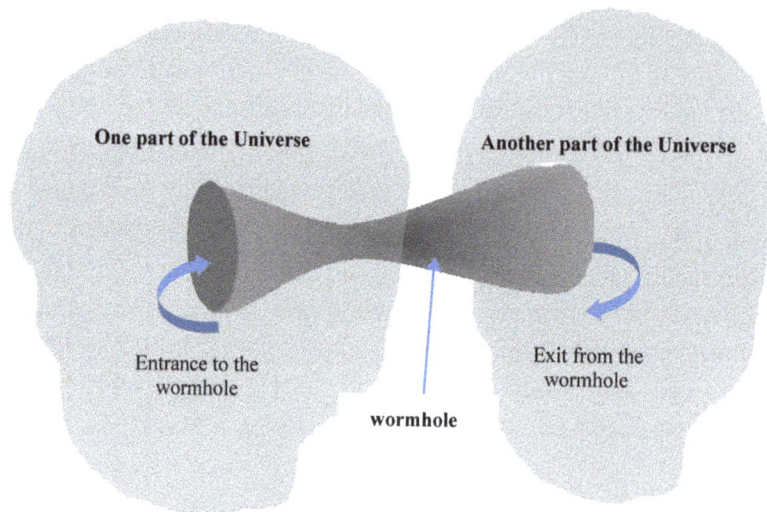

Fig. A.14 Wormhole.

Einstein's theory allows for the possibility of the existence of *wormholes*. However, whether they exist or not is pure conjecture at present. But if they do, and if a civilisation has the capability of finding or creating them, and controlling them, they can be used as a short cut for travel between two distant places in our Universe and perhaps between two points in time. [Note: nothing in our science allows travel to the past before the time that the time-machine used in the travel was created].

White holes

Strangely, Einstein's theories also allow for *white holes*. White holes are the opposite of black holes. Unlike black holes where matter and light enter the hole never to be seen again, white holes would be regions in space from which matter and light pour out, but which themselves cannot be entered!

Appendix A. Riding on a light beam…

Do white holes exist (as we once wondered about the black holes)? What would be feeding the white holes? Would the black and white holes be connected? Would the matter going into the black holes be coming out of the white holes? The mind boggles.

What have we learnt?

We started by talking about Albert Einstein, and about his genius. We discussed the ground-breaking theories he produced in a set of papers: in 1905 on his *theory of special relativity*, and in 1916 on his *theory of general relativity* that extended the Special theory and showed the effects of gravity on light, space, and time. We saw that Einstein would think through his ideas in thought experiments, using analogies that made the problems more understandable.

We went through some of these thought experiments and saw the strange results they gave about the relationship between light, time, and space. We considered the implications of the ultimate speed, that of light. We reflected on the strange effects that occur when objects travel at relativistic speeds (speeds near that of light): their mass increases, their time slows, their lifetimes lengthen, their lengths reduce, the distances they travel reduce, as measured by stationary observers. We considered the possibility that such speeds give, in theory at least, of time travel into the future (but not the past).

We have seen that gravity is interchangeable with acceleration. If we were in a room unable to look out, we would not be able to tell whether we were feeling the effects of gravity while we were stationary or whether the room that we were in was in fact accelerating in space.

We have also seen that gravity can affect time itself, and that clocks run slower in regions of higher gravity.

We further found that gravity can distort spacetime so that the path of a light beam is no longer straight but bent towards the massive object distorting the spacetime. We have seen that this can cause stars to appear to have shifted when their light passes near some other massive object.

We saw how gravity can stretch a light's wavelength (making it 'redder'), when it is travelling away from a massive body and compress it (making it 'bluer'), when it is travelling towards it.

We noted that Einstein's work led to the identification of the ultimate effect of gravity, the creation of a black hole.

Finally, in keeping with our scientific approach, we saw that Einstein's work, as all serious scientific effort, is underpinned by experiments. We reviewed a few experiments that support the somewhat bizarre conclusions to which Einstein's theories lead us. These experiments range from the measurement of the speed of light, which demonstrated that the speed was always the same, as predicted, to the bending of starlight by gravity and the increase in lifetimes of muons, the particles created high above the Earth by collisions between cosmic rays and the atmospheric molecules. We saw that CERN's Large Hadron Collider had been important in many of these experiments.

The conclusions of Einstein's predictions have been extensively studied, analysed, and experimented on. All areas where experiments could be carried out have come through with flying colours. There remain very few far-out predictions of the theory, such as white holes and wormholes, which have not yet been investigated simply because we do not yet have the understanding nor sensitive enough apparatus. One of the predictions shown to be correct recently, is the existence of gravitational waves, which are caused by immense collisions way out in space. The sensitivity of experiments is continuously being extended, and new techniques are evolving. The search carries on.

Einstein showed that the laws of physics and the speed of light were the same for all observers who were moving uniformly, without acceleration, in space.

This simple statement is extremely significant. The first part, about the laws of physics, states that we must be careful of the conditions under which we measure any object to get a clear picture of what is happening and how the laws apply. We have seen that acceleration and gravity can be considered the same, and the effect that gravity can have on space and time itself. Thus, when we compare two events, we need to make sure that the events are taking place when the two objects are moving uniformly and are not subject to acceleration between themselves.

The second bit about speed of light is one we have mentioned during our thought experiments and earlier. Before Einstein, scientists were worried because light did not appear to obey the laws of motion that Newton had developed. Experiments showed that whichever way we measured it the speed of light was always the same. But no one really knew why. Other scientists found that they could resolve some of the peculiar results they were getting, but only by assuming strange things in their experiments. Time and sizes of objects seemed to change depending on how you measured them and how fast things were travelling. No one had an explanation until Einstein's Special theory came along in 1905.

Einstein showed that the speed of light was a constant, the same under all circumstances and for whosoever observed it. It did not matter which direction light was going, or whether the observer was moving towards it or away from it. It was always the same.

He further showed that some of the bizarre assumptions that had to be made about objects and distances shrinking in size, and time depending on the observers' relative speeds, were in fact correct. Space and time were not independent from each other. In other words, their values depended on the observer who is measuring them. But light is different: its speed is always the same.

Appendix A. Riding on a light beam…

Later, a German mathematician called Hermann Minkowski (1864-1909), proposed that we existed in a four-dimensional Universe, three dimensions of space and one of time, and suggested that space and time were linked together inseparably. Einstein developed this concept further when he worked on his General theory, which was published in 1916. He showed the inter-relationship of space and time and that they should be considered together as *spacetime*.

Einstein's *theory of general relativity* is highly mathematical and complex. It covers the concept of spacetime and shows the effects and implications of gravity. Einstein adopted a quite different approach to gravity from that of Newton. He no longer talked about the force of gravity but showed that gravity is a result of the interaction between spacetime and matter. Earlier we discussed the famous analogy of spacetime represented by a sheet of rubber (Fig. A.8), that is often used to explain the concept that matter warps spacetime.

Massive objects can thus distort spacetime in their locality. Gravity can be considered as a curvature of spacetime by matter, in other words, gravity is the effect that matter has on spacetime. Distorted spacetime will affect light as well as matter, since it is space itself that has been distorted. Hence Einstein's prediction that gravity would bend light by distorting spacetime.

The implications of this can be seen in the extreme gravitational field caused by a black hole. Light passing by a black hole will be bent, more and more, the nearer it approaches the hole. When it skims past the event horizon of a black hole, it will be curved to such as extent, that it is captured and keeps encircling ('orbiting') the black hole. Light that is within a black hole simply cannot escape the clutches of gravity.

Einstein's work extended into the effects of gravity on spacetime, and into exotic topics such as the curvature of space, which we met in Chapter 7. His equations of general relativity were complex, but solutions were produced as we have discussed. Specifically, we saw that an exact solution was produced by four scientists: Friedmann, Lemaitre, Robertson, and Walker.

Einstein was, without doubt, one of the leading thinkers of all time. His theories have and will continue to influence the direction of human scientific progress for generations to come.

Appendix B

The strange world Alice would see…

if she sat on an atom

In the previous appendix we glibly assumed that you could reduce yourself down to a minute size and jump on to a photon. But this really would not be possible. You are made of matter which is made of molecules which are made of atoms. So, if you were that tiny, each particle you were made of would need to shrink. You could no longer be made up of the atoms and molecules we know of, but of strange new tiny particles that would need to come into existence out of nowhere. So, it would appear that no one, not even Alice, could ever become that small.

But there may be a way. We saw earlier that according to Einstein's theory, the faster that something travelled, the more it shrank in size, at least so it appeared to an outside observer. So, if you travelled extremely fast, you became extremely small, rather like Alice. However, this creates another problem.

You are made of matter and have mass. Light photons on the other hand are energy, and do not have mass. Therefore, they whiz along at the speed of light. You need energy to travel fast and, as we saw in the last chapter, the faster you travelled the more massive you became. Travelling at the speed of light would make you infinitely massive, and you would need an infinite amount of energy to travel at that speed, which is, of course, impossible. This all means that you cannot travel at the speed of light after all.

But even if we cannot travel as fast as light, we can imagine whatever we like. Therefore, jumping onto a photon is quite acceptable as a thought experiment. So, let us now imagine that Alice drinks a magic potion and does manage to shrink herself down to a tiny, tiny size without becoming infinitely heavy, and is able to perch on an atom. *If* Alice could do this, what sort of world would she find at those minute sizes? This is what this chapter is all about.

Come along as we journey into the strange world of the *quanta*, the place of the tiny electrons and photons, where things can exist in many places at the same time, where particles can pass through solid barriers, where things can have mass as well as be a bundle of energy all at the same time, where events change simply by being observed. It is a world that was suggested by Einstein's theories, but it is a world so peculiar that Einstein refused to accept that it could be so. However, it is also a world that is backed by the most complete theory we know, *Quantum Mechanics*, and supported by extensive experimental results.

In Physics, *quanta* refer to a minimum discrete packet of energy or matter (such as photons, electrons, atoms) that can be involved in an interaction. This is the world that is covered by the science of *Quantum Mechanics*.

The theory behind quantum mechanics was developed in Europe over the early years of the 20th century. It started with Einstein's important 1905 papers on photo-electricity in which he showed that light particles can produce electricity when they strike certain materials, for which he received the Nobel Prize in 1921. (Interestingly, Einstein did not get a prize for his work on relativity, perhaps they were waiting for all his predictions to be proved!). Later work by several scientists, in particular the French, Louis de Broglie (1892-1987), the German, Werner Heisenberg (1901–1976), the Austrian, Erwin Schrodinger (1887-1961), and the English, Paul Dirac (1902-1984), amongst others, developed the early ideas into the extensive Quantum Mechanics theory we have today.

Here we shall look at some interesting aspects of the theory that these and other scientists put together, and not at the detailed and complex theoretical basis of the theory itself.

Appendix B. The strange world Alice would see…

The world of the quanta

We will not be doing a detailed study of the fascinating, but complex, subject of quantum mechanics. Rather, we will look at some of the aspects of the strange world it describes. I want you to remember that what we will be talking about is not fiction, or the ramblings of a lunatic, but is provable, by theory, mathematics, and experiment. Quantum mechanics is the most precise and provable theory we have.

God does play dice

Quantum theory deals with probabilities. That is, it deals with the chances of things happening and the rates at which things happen, rather than with absolutes. For example, we know from the study of nuclear physics that radioactive elements, such as uranium and radium, lose mass (decay) at specific rates due to the loss over time of certain atomic particles from their nuclei. This shower of particles is the nuclear radiation which can be so harmful to people's health. The decay happens spontaneously and at random intervals. In other words, it is not possible to predict exactly when the next particle, or any specific particle, will decay. However, if we measure the mass of the element sample that we are using, and then measure it again after some time, we find that loss of mass over this time can be predicted *exactly*. The time it takes for any sample of an element to lose half its mass due to radiation is called its *half-life*.

Each element has its own half-life; some highly radioactive elements have a short half-life and some an extremely long one. For example, the half-life of tellurium-128, an isotope* of tellurium is 2.4×10^{24} years, which you will remember is 24 followed by 23 zeros. You will recall that our Universe is only 13.8×10^{9} years old. So, tellurium-128's half-life is more than 10^{14} (hundred trillion) *times* the current age of our Universe! At the other end of the scale nitrogen-10 has a half-life of 2×10^{-22} seconds.

Apart from showing that *everything* decays over time, the interesting point of the discussion about half-life is that while we cannot predict by any means which specific atom is going to decay, we can precisely predict what a collection of atoms is going to do. This can be demonstrated no matter how many times the experiment is repeated. It turns out that the workings of nature really *are* based on probabilities and chances rather than absolutes.

Heisenberg's uncertainty principle

We noted above that Heisenberg was an important pioneer of the work on quantum mechanics. In 1927, he defined what is called the *Heisenberg uncertainty principle*.

---

* An isotope is a form of an element which has the same number of protons as all the other forms of that element, but a different number of neutrons.

This principle states that we can *never* measure exactly both the position and velocity (speed in a specific direction) of any object at the same time. In other words, if we know how fast the object is travelling, we can never know exactly where it is, and vice versa.

This seems strange, since we frequently measure the speed of objects such as a car, while knowing exactly where it is, especially when we are in it. This is because, for large objects, the uncertainties are extremely small compared with the measurements of speed and position.

The uncertainties become important when the object is the size of the atomic world. Here is an analogy to show how this can be so.

Suppose you have a microscope which is powerful enough to allow you to see an electron whose position you now want to measure. The only way we can see any object is by the photons which hit the object and reflect off it into our eyes. However, as a photon of light strikes the electron, it gives the electron a shove, and by the time the light reaches your eye, the electron is no longer where it was. The result is that there is an error in the position measurement.

Heisenberg produced a specific, mathematical equation that shows the relationship between the uncertainties of the momentum (which is velocity multiplied by mass) and the position of any object. This equation says that the more certain we are of the momentum of an object, the more uncertain we are of its position, and vice versa.

Light is both a wave as well as being made up of particles

You must have dropped a pebble in a pond at some time, and seen the circular ripples move away from the point where the pebble hit the water. This is what happens to all waves, whether they are in a pond, sea, or air, and whether they travel in water as water waves, in air as sound waves, or in space as electromagnetic light waves. Waves radiate away from their point of origin.

Now imagine you have put a metal plate with a slit cut in it in the path of the expanding circle of waves resulting from the pebble you dropped into the pond [Fig. B.1 (a)]. What will now happen is that on the side of the plate facing away from the incoming waves, a new set of waves will form, radiating away from the metal plate with their centre at the middle of the slit where the metal meets the water. This is a quite an easy experiment to perform and can also be done in a bathtub. This happens to all waves, in liquid, solids, air or space.

Next imagine that you have a metal plate with two parallel slits cut in it. Again, you place it in the water as we did with the single-slit plate [Fig. B.1 (b)]. What do you think will now happen?

Appendix B. The strange world Alice would see...

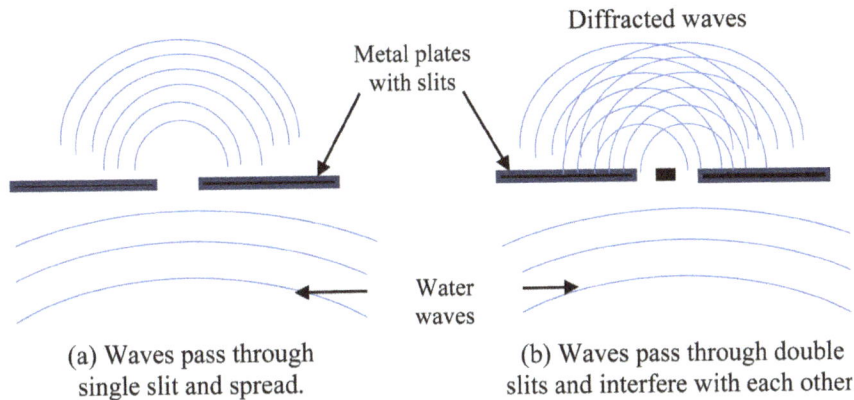

Fig. B.1 Water waves meeting barriers with slits.

Well, what happens is that we get two new waves forming, each having the middle of *their* slit, where the wave escapes, as its centre [Fig. B.1 (b)].

Now for the interesting bit. The two sets of waves formed by the two slits will cross each other (or scientifically, the two sets of waves will interfere with each other) as they spread. If we take careful photographs of the waves, we will find that at some points, the resulting waves are higher than normal, and at others they are lower than normal. At some points, the water will be flat. This pattern is called an interference or diffraction pattern, and all waves behave in this way. It is easier to show what is happening with a diagram [Fig. B.2].

The diagram shows two waves, which are moving along. Let us imagine the two waves cross each other as shown.

Depending on the frequency of the waves and how they cross, there will be points where the peaks of the two waves occur at the same place. When this happens, the result will be an even higher (reinforced) peak. This is called *constructive interference*. If the two waves cross where one of the waves has a peak at a point where the other wave has a trough, the peaks will be reduced and, if equal in size and frequency, will cancel each other out resulting in a wave of zero height. This is called *destructive interference*.

Now, interesting as this story is, it is perfectly okay for you to ask, "So what?" The answer to this question is one of the most fascinating ones in science. Let me explain.

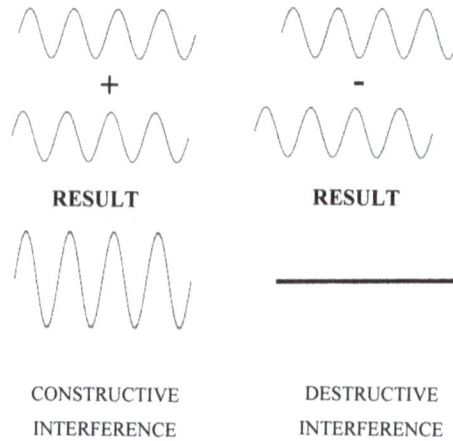

+

-

RESULT

RESULT

CONSTRUCTIVE
INTERFERENCE

DESTRUCTIVE
INTERFERENCE

Fig. B.2 Interference caused by two crossing waves
can be constructive (+), destructive (-) or in-between.

We learnt elsewhere in the book that light is made of waves. We would therefore expect that if we were to conduct a similar 'two-slit' experiment with light, we would see similar evidence of the light beams with peaks and zero wave heights as we obtained for water waves. We would be right.

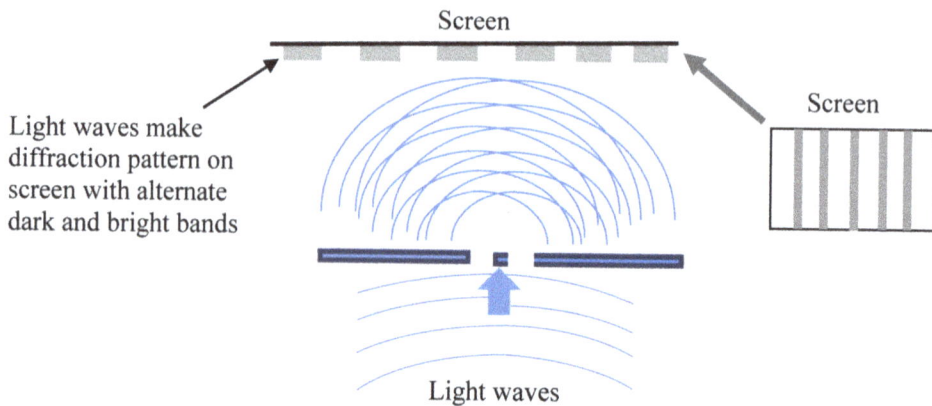

Screen

Screen

Light waves make diffraction pattern on screen with alternate dark and bright bands

Light waves

Fig. B.3 Interference pattern caused by two crossing light beams.

We can perform the experiment as shown in Fig. B.3.

Appendix B. The strange world Alice would see…

When we do this with a beam of light and two slits in, say, an opaque cardboard sheet, and with a screen (cardboard or wall) on the other side, we indeed find that we get a pattern of bright and dark lines on the screen. The bright lines correspond to the places where the peaks of the light waves coincide, and reinforce each other, and the dark lines correspond to the places where the peaks and troughs of the waves cancel each other. This was demonstrated by Robert Young, a British scientist and medical physician, in 1801.

Robert Young (1773-1829) was the eldest of ten children. He was a polymath, with many skills in the sciences, as well as being a physician and a linguist. He was especially knowledgeable in light and mechanics, as well as medicine, music, and the study of ancient Egypt. He is said to have understood English, Latin, Greek, French, Italian, Hebrew, German, Aramaic, Arabic and Persian. Altogether, he was a remarkable man.

Now back to the experiment and the fascinating developments that followed.

More than a century later in 1927, two American physicists, Clinton Davisson (1881-1958) and Lester Germer (1896-1971), performed an experiment in which they repeated Young's experiment, but with electrons rather than light. They fired electrons at a 2-slit barrier set up just as in Young's experiment. They found that *they obtained an interference pattern* just like Young's (using a screen sensitive to electrons, of course). This was amazing because electrons were supposed to be particles with mass, and yet here they were behaving like waves.

Davisson was awarded the 1937 Nobel Prize in Physics for discovering electron diffraction (the wave nature of electrons) during the famous experiment.

In the late 19th century, when Maxwell was doing his pioneering work on electromagnetic fields, light radiation was thought to be electromagnetic waves and matter to be made of particles. Then in 1905 Einstein published a paper on the fact that when light was shone on some materials, electrons were ejected and could be measured as an electric current. In other words, he showed that light itself was behaving as if it was made up of particles called photons which had the energy to knock electrons out of atoms. His work on the photoelectric effect won Einstein the 1921 Nobel Prize for Physics. (Incidentally, the photoelectric effect is the way solar panels, that you see on many roofs, work to produce electricity from sunlight).

So, not only does matter behave as if it were both waves and particles, but light also behaves as if it were both waves and particles.

This confirmed a hypothesis that had been put forward in 1924 by Louis de Broglie, the renowned French physicist, which stated that electromagnetic energy (light) and particles such as electrons would both behave either like a wave or a particle depending on how they were measured. In other words, electrons, and light both behave as if they are *both* waves *and* particles and show themselves as one or the other depending how they are being observed. This is called the *wave-particle duality* and is a fundamental aspect of quantum mechanics.

Another point that came from de Broglie's work was that while the energy of particles depended on the velocity (speed in any given direction) with which the particles were moving, that of light depended on the frequency of the waves (their speed being constant). The faster the particles move the higher is their energy, and the higher the frequency of the waves the higher is their energy.

From a quantum mechanics point of view, the amazing story does not end with the pattern that forms from the slits with both light and electrons. Suppose we now reduce the strength of the electron beam so that ultimately only one electron is released at a time. We would expect that the electron would go through one of the slits or hit the barrier and be absorbed. The electron that lands on the screen shows up as a spot of light. Eventually, we would expect that just two bands are formed, parallel to the slits, depending on which slit the electron passes through. However, what we find is that *the final pattern is the same as the diffraction pattern that formed when a beam of electrons was fired at the two slits*. We argued then that the pattern resulted from the two waves interfering with each other. But if an electron is going through one slit on its own, what is going through the second slit to interfere with it? The conclusion must be that the electron is behaving like a wave and going through both slits at the same time and interfering with itself!

The strangeness does not stop even there. Supposing we fire a stream of electrons at the slits as before. We will get the interference pattern we have discussed. Now, suppose we have rigged up a detector that counts each time an electron shoots through one of the slits. We would expect the detector to count half the number of electrons that were fired, and for the interference pattern to show up as before. The remarkable result is that while the count is half the total, as we expected, only two bands are seen, as if the electrons are behaving only like particles. When we switch the detector off the electrons go back to demonstrating their wave-like behaviour and the interference pattern re-emerges. How do the electrons know whether they are being observed or not, and which way to behave?

Remarkably, the same effects are observed with a beam of light replacing the electron stream, and with photons in place of electrons. Energy and matter behave like each other at the quanta-scale.

This is the mysterious quantum world that is all around us but hidden from us because of our size.

We therefore need to ask what is a particle in this strange world?

Appendix B. The strange world Alice would see...

Particles that span the Universe

If particles are waves, they must have a 'wave equation' that mathematically specifies the shape, frequency, and position of the wave, just as scientists can specify an equation for a light-wave. The wave equation for a particle was defined by Erwin Schrodinger, an Austrian physicist who won the Nobel Prize in Physics in 1933 for his work on what is now known as *Schrodinger's equation.*

Schrodinger's equation defines a particle as a system operating through space and time. It specifies, for example, the likelihood of a particle being found at a particular location.

The wave equation for a particle is called its wave function; it specifies the location of the particle at a point of time in terms of probabilities, that is, in terms of the likelihood that a particle will be found at a particular location. The peak of the wave indicates the most likely location of the particle, but the shape of the wave function, which does not hit zero till infinity, does not rule out the tiny possibility of finding the particle elsewhere *in the Universe* [Fig. B.4]. Thus, our particle has a chance of being anywhere and everywhere in the Universe, until it is observed at a particular point at which instant its wave function collapses, and the particle exists just as a physical particle solely at that point.

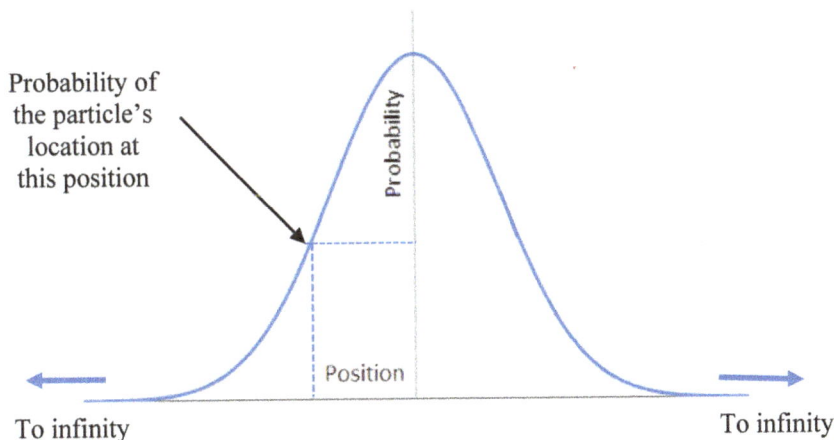

Fig. B.4 Wave function of a particle.
The curve gives the likelihood of the particle being found at any given position.
The ends of the curve stretch out to infinity, thus the particle has the possibility
'of being found' anywhere in the Universe.

The dead and alive cat

Schrodinger defined a thought experiment in 1935 which shows the strange implications of quantum mechanics when applied to the normal world so familiar to us.

Imagine that there is a box which contains a cat, a pellet of a radioactive element and an arrangement such that when an atom of the radioactive element decays it results in the release of a hammer that breaks a container releasing a poisonous gas which instantly kills the cat. However, no one knows when that will happen since the atom decay is completely random. It could happen within a second, an hour or indeed not within the period of the experiment. So, the question is, at any instant, "Is the cat dead or alive?" We cannot know the answer until the box is opened and the cat seen. The Schrodinger's equation for the box system which contains the cat allows for all possibilities. The only way we can find out what has happened is by opening the box and observing. Until then the cat is both alive and dead. At the point we observe the situation, the equation collapses, and we discover the state of the cat.

But how do we know whether the strange things we have been speaking of are so, and how relevant are they for normal life? Let me give you an example to answer these questions.

Tunnelling: how our electronic devices work

We are all familiar with TVs, radios, computers, smart phones, and other such electronic devices. They are now part of our everyday life. We use them for work, play and learning. They provide us with information, tools to use and games to play. But how do these devices work? You may have heard of things called 'chips' which are small pieces of silicon on which are etched electronic components such as transistors, resistors, capacitors. Some chips have memory circuits on which to store the work you have done, stories you may have written or photos that you may have taken. Today we cannot imagine life without them.

When I was first working as an electronics engineer, these components were much larger than they are now. A transistor could be as big as a pea, whereas today you can put billions of them into that space.

These electronic devices work by switching electronic signals around a circuit, amplifying the signals, adding, subtracting, and combining them, storing the information in a memory, and displaying the result on a screen. An essential requirement for a transistor is to be able to control and manage the electronic signal. The electronic signal is made up of electrons moving along the circuit. These electrons can be stopped or diverted by applying other signals to the transistor, like the way that trains are controlled by signals along their tracks.

Now for the interesting bit. When we stop an electronic current in a transistor, we do it by putting a barrier in its path. An electron has a negative (-) charge, and since like charges repel each other (as do like poles of a magnet), a negative charge will act as a barrier for an electron. This barrier acts as if we have built a wall across the road, down which the current (electron) is

Appendix B. The strange world Alice would see…

travelling. Our normal experience tells us that this barrier should stop the flow of electrons, just as we would expect a wall to stop a car. However, we find that the not all electrons are stopped, some manage to go through the barrier and are to be found on the other side. It is as if some cars manage to squeeze *through* the wall. In fact, this strange effect, which is called tunnelling, is essential for making the transistor work. But how does this happen?

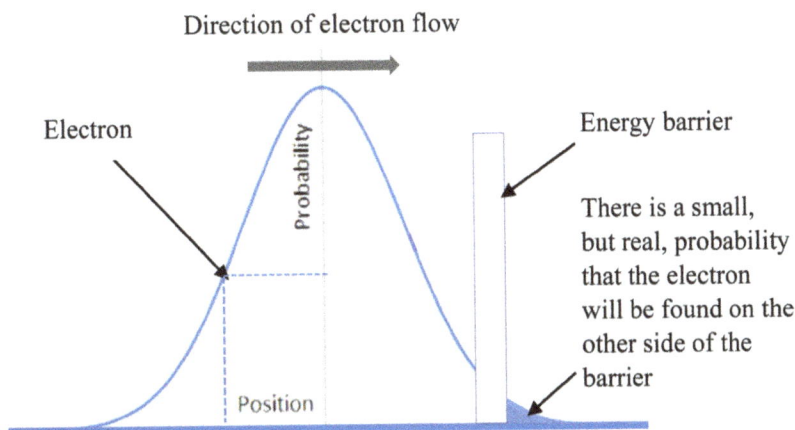

Fig. B.5 Electrons (electric current) can 'leak' across energy barriers.

The curve shows where the electron may be found.
The height of the curve at a point gives the likelihood (probability)
of that electron being found at that specific position.

We have seen that the electron can be described either as a particle, or in quantum physics terms, as a wave function. Quantum theory says that the wave function is spread over space with a peak showing where the electron is most likely to be found. If we put the wave function of the electron against a barrier wall as shown Fig. B.5, we find that there is a tiny bit of the wave function (shaded) that is on the other side of the barrier. In other words, there is a small but finite chance that the electron is to be found on the other side of the wall. This is in fact what happens. Some of the electrons 'leak' across the barrier because their wave function says they can. The wave function also tells us the chance of this happening and thus how many of the electrons would be found on the other side. This is exactly what occurs, showing us that the quantum world is as bizarre as the quantum theory says it is.

Weird, or what?

Quantum tunnelling is needed for the stars and galaxies to exist

We learnt earlier in the book that atoms are made of a nucleus, which contains two types of particles: protons which are positively charged and neutrons which do not have a charge but have a mass which is virtually the same as that of the proton. Whirling around the nucleus, is a cloud of electrons which are negatively charged. Each atom has an equal number of protons and electrons, so the positive and negative charges balance each other.

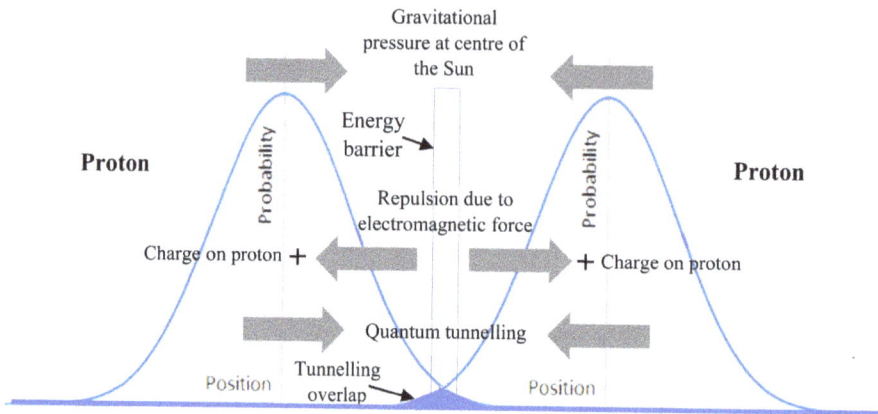

Fig. B.6 Quantum tunnelling enables the Sun to generate its energy.

The simplest atom, hydrogen, has just one proton at the nucleus and one electron spinning around it. The next element, helium, has two protons and two neutrons at the nucleus and two electrons in orbit around it. More complex atoms have even more protons and neutrons at the centre, with the number of electrons in orbit equal to the number of positively charged protons.

We know that the stars get their energy from converting hydrogen into helium. To do this, four hydrogen nuclei (positively charged protons) need to come together to create a helium nucleus of two protons and two neutrons. But here is the problem: the protons being of the same charge will repel each other due to the electromagnetic force, which we have seen is one of the four fundamental forces of nature. So how can they be forced to join up?

Well, there is an enormous gravitational pressure at the centre of stars, due to the mass of their gas bearing down on the ions (atoms which have had their electrons stripped away) there, forcing them together. But there is also another factor at play. This is the tunnelling effect. The wave functions of the two particles can overlap and so enable the protons to cross the barrier put up by repulsion of the electromagnetic force, at a much lower energy level than would otherwise be required for combination due to gravity alone [Fig. B.6].

Appendix B. The strange world Alice would see…

Quantum tunnelling operating on the nuclei of the particles in the centre of the Sun enables it to generate its energy and shine at core temperatures of 15 million C. Without these quantum effects, temperatures of about a hundred million C would be needed for the protons to combine to produce helium. Without quantum tunnelling, our Sun would not shine and there would be no life on Earth.

Spooky action at a distance: Quantum entanglement

Quantum theory says that it is possible for the wave functions of two particles to get "entangled", so that the two behave as if they are one particle, with a common wave function, *wherever the two particles may have been moved to after their entanglement, even to 'the opposite ends of the Universe'*. Einstein called this "spooky action at a distance".

The statement above seems to imply that information can pass between two objects faster than the speed of light. Now we know from Einstein's *theory of special relativity* that nothing, not even information, can travel faster than the speed of light. So, what is happening?

Particles, including photons and electrons, can be described by their properties, such as mass, charge, momentum. One such property is called *spin*. This does not mean that the particles are spinning, but it is a property we can measure. Spin can only be in one direction or its opposite (spin up or down). Like other quantum properties the spin of a particle only becomes known when it is observed or measured.

When the spin of two free particles is measured, there is no relationship between their individual spins. If one has an up spin, the other can have either an up or a down spin randomly. If the particles are created at the same time though, they can be 'entangled' with each other (how this is achieved is beyond this book). Remarkably, in the case of entangled particles, each particle's spin when measured will *always* be found to be the *opposite* of the other particle's spin.

The particles behave as if each knows what the other's spin will be at any instant, even if the other particle is, 'at the other end of the Universe'. All attempts to disprove entanglement have failed. But *how* does the second particle know what was the final spin of the first without information being passed across? Spooky, indeed.

Einstein was not happy. To get around the problem, he reasoned that all particles must be carrying 'hidden information', a plan as to what their spin will be when it is measured, and that on entanglement, the plans of the particles get set up such that the measurement on the second particle always produces the opposite result to the first. This theory was tested, and it was proven that there is no hidden information being carried by the particles.

It is now accepted that the entangled particles somehow produce the result that they do. Entanglement works, but we do not know how. What we know though is that, under our present knowledge, it must do so without breaking the rule of *information* not travelling faster than the speed of light. When we use an entangled particle for communication, it must be observed and its entangled state collapses immediately.

Quantum mechanics in general, and entanglement specifically, is proving extremely valuable in areas such as quantum computers which are now coming into use for solving complex problems. These computers which are based on quantum principles, work by breaking the problem into multiple parts and solving these parts in parallel. Exactly how they do this is another mystery. It is as if given a problem of finding a way out of a maze, the quantum computer tries out all paths in parallel, at the same time and instantly picks out the correct one. Entanglement is also being used in cryptography, to ensure extremely secure communications which are safe from hacking.

The distance between entangled particles has steadily increased since the initial experiments in 1997 in Austria. In 2016, Chinese scientists succeeded in sending a quantum lab up into space. There two photons were entangled and then beamed down to two laboratories 1000 km apart on the Earth. Remarkably, they remained entangled.

Entanglement is also being used in quantum teleportation. This is not the beaming of people, as in Star Trek, but transmitting the state of a particle, its spin, etc., onto another particle in a remote location. In 2017, the Chinese successfully teleported the properties of a photon on Earth to another photon in orbit.

Conclusions

We have seen a glimpse of how our Universe works at the tiny quantum scales. All these strange effects are backed up by a very robust theory, and experimentation. But for physicists, one critical problem remains. We have Einstein's *theory of general relativity* giving us a precise knowledge of the Universe at the large scales, and we have the *quantum theory* giving us a very precise specification of the Universe at the atomic scales, but we have not been able to marry the two.

Einstein's *theory of general relativity*, if used at the atomic scale produces nonsense, and *quantum theory* does not operate at the large scale. There is a great deal of work being undertaken to resolve this conundrum, and to produce a *Theory of Everything* which will cover both the huge and the tiny. There are some promising leads being followed, and theories being worked on, such as the string theory, M-theory, and quantum gravity some of which we have come across earlier in the book.

Hopefully, we will celebrate discovering the ultimate theory before long.

Appendix C

Three of my favourite scientists...

Albert Einstein, Paul Dirac, and Richard Feynman

Before concluding, I want to introduce you to the lives of three of my favourite scientists who worked at the cutting-edge in the field of theoretical physics. One, Einstein, upended the scientific study of the huge with his work on *Relativity*, the other two helped to create the new science of the tiny: *Quantum Mechanics*. Without these three, progress in our understanding of the natural world would not only have been much slower and our knowledge much poorer, but our lives would also have been duller.

All were geniuses, undoubtedly, easily amongst the greatest intellects of any time. But they were also fascinating characters, though each quite different from the others.

Albert Einstein

Albert Einstein [Fig. 1.10] was born on 14 March 1879 in Ulm, Germany, into a middle-class, German, Jewish family. He was a good student, but his creative, imaginative mind did not always fit in with the rigidity and strict disciplinarian approach of the schools he attended. His education was patchy. Eventually, he left Germany and settled in Switzerland where his family had migrated. After graduation, he spent two years unable to find a suitable teaching post that he was seeking. Finally, in 1903, with help from a friend's father, he got a job at the Swiss Patent Office. Imagine the great physicist-to-be as a patent clerk!

The job had one great advantage for the young Albert. It gave him time to think. And thinking is what he did best of all. He was fond of conducting his thought experiments about a problem in which he imagined various '*what if*' scenarios, to consider what would happen to the issue he was studying under the different imagined circumstances. As we have seen in Appendix A above, one famous thought experiment was when he wondered what he would see if he was able to ride on a light beam.

Amazingly, in 1905, his *annus mirabilis* (auspicious year), at the age of 26 years, within two years of joining the Patent Office, he had produced some of his most profound work. He published three major papers. One, on the *photoelectric effect* showed how light can interact with atoms and electrons to generate electricity, as occurs in today's solar panels. This work became the initial step into the quantum mechanics revolution. Another paper, on a subject called *Brownian motion*, helped to demonstrate that atoms and molecules exist. The final one was on the *theory of special relativity*, special because it dealt with relativity for only certain specific cases. In this work he brought together space and time into what is known as *spacetime* and showed what happens when objects approach the speed of light.

In 1916, Einstein extended his work on special relativity into the *theory of general relativity*, which brought gravity into play, together with light, space, and time, and showed how gravity affects them and how this effect can explain some of the anomalies of physics. This work led to his insight on the possible existence of black holes (though Einstein thought these were too bizarre to exist) and proved that mass and energy were the same thing. This in its turn led to nuclear power, and the atom bomb. It was altogether a breath-taking achievement.

Surprisingly, when Einstein won a Nobel Prize (in 1921) it was for his work on the photoelectric effect, and not for relativity. Possibly it was because some of the deciding committee did not understand the subject, but to be charitable let us say that his theories on relativity at that time could still be subject of debate.

Appendix C. Three of my favourite scientists…

In 1933 Einstein migrated to the USA to escape Hitler's clutches. He joined the Princeton University's Institute for Advanced Study. During the war years, afraid that Nazi Germany would develop the atom bomb first, he worked on this for the USA.

He regretted the destruction that the atom bomb caused when it was dropped on Japan and worked hard to bring it under control. He became involved in the US civil rights movement. In his field of physics, he tried to bring together the new quantum theory and relativity. Einstein found it difficult to resolve quantum mechanics' bizarre (but proven) conclusions with his logic. He famously said that "God does not play dice", when faced with the consequence of the new science that nature works on probabilities rather than on certainties. In his later years he also spent time on far out topics such as worm holes, time travel and black holes.

Einstein was 76 when he died in Princeton, USA, on 18 April 1955.

Paul Dirac

Paul Dirac was born in Bristol, UK on 8 August 1902. His father, Charles, had emigrated from Switzerland; his mother, Florence, was English from Cornwall. He was the middle of three siblings, with an elder brother, Felix, and younger sister, Betty. Charles Dirac and the three children became UK citizens in 1919.

Dirac grew up in a strict household; he was not close to his authoritarian father and became reclusive and withdrawn. Tragedy struck the family when Felix committed suicide in 1925. It is revealing that Paul Dirac said later that, till he saw his parents' anguish he had not realised that parents were supposed to care for their children. He was a very strange man indeed.

However, though he lacked empathy, he was gifted in many other ways. He sailed through his education in engineering and mathematics at Cambridge and other institutions with a series of scholarships and honours. He became interested in relativity and then quantum mechanics.

Dirac's work in the field of theoretical physics is littered with honours and prizes. In 1928, he developed a famous equation, now called the Dirac equation, which combined quantum theory and Einstein's special relativity to define how an electron behaves when travelling at relativistic (near-light) speeds. This equation predicted in 1933, the existence of the antimatter* particle for an electron, called an anti-electron or positron. This prediction was made by Dirac well before the first antimatter atoms of hydrogen were demonstrated to exist by being created in the Large Electron-Positron Collider, the predecessor of the Large Hadron Collider, at CERN on 4 January 1996.

---

* Note: antimatter particles exist for each fundamental particle of nature. When a particle meets its antimatter twin, they are both annihilated in a burst of energy.

Paul Dirac was awarded the Nobel Prize in Physics in 1933 for his work on atomic theory, which he shared with Erwin Schrodinger, of the dead-or-alive-cat-in-the-box fame.

Dirac's life is described in a biography by Graham Farmelo aptly called "The Strangest Man – the hidden life of Paul Dirac, quantum genius".

Paul Dirac did meet Richard Feynman, the third of my physicist choices. The story goes that when they met at a conference, he said to Feynman, after a long period of silence, "I have an equation, do you have one too?"

Dirac died on 20 October 1984 in Tallahassee, Florida, USA.

Richard Feynman

Richard Feynman was a completely different character to Einstein and Dirac. He was a larger-than-life American, who was a humourist, a talented bongo player, a brilliant teacher, an amateur safe breaker, an inventor, and an outstanding physicist.

Feynman was born on 11 May 1918 in Queens, New York City, USA, to Melville Feynman, a sales manager and Lucille. He had a younger brother, Henry, who died soon after birth. Henry was followed by a sister, Joan, who was 9 years younger than Richard. Joan with her brother's backing developed her interests in astronomy and became an astrophysicist.

Feynman was encouraged by his father to be enquiring and not to take things at their face value. From his mother, he inherited his sense of humour. He was fascinated by engineering, especially radios. He was very inventive, and when he was eleven or twelve, he set up a laboratory at home for his experiments. He wrote about this early phase in his bestselling autobiography *"Surely You're Joking, Mr. Feynman!"*; which is a fun read as well as informative. He related how when he was 11 or 12 years old, he set up a contraption using bulbs and switches and fuses to make a controllable lamp bank. [Personal note: This story appealed to me greatly as I had attempted something similar in my youth. There was a significant difference though; Feynman had made sure that his circuit fuse time was shorter than his home's fuse, and he put controls in his circuit to manage his lamp bank. Mine was simply a bank of bulbs, which when switched on blew the fuse of the whole of our house, though with a satisfying bang! – Note: this is another one of my 'must-not-to-tried-without-supervision experiment' warnings for young budding scientists.]

By the time he was 15, Feynman had taught himself advanced mathematics. He went on to study mathematics and then physics at MIT (the Massachusetts Institute of Technology), then to Princeton University in 1939 for his graduate studies. It is said that he obtained a perfect score in his Princeton entrance exam in physics, a unique achievement at the time. He was a contemporary, at Princeton of scientists such as Einstein and Pauli. He obtained his doctorate in 1942.

Appendix C. Three of my favourite scientists…

One of Feynman's major achievements was developing a diagrammatic way to represent the interactions of sub-atomic particles. These diagrams, which became known as Feynman diagrams, made this complex subject much easier to represent and discuss. He was by now one of the best-known physicists in the world and recognised as one of the greatest.

Feynman went on to assist the war effort, working on the Manhattan Project to develop the atom bomb, which helped end World War II. Later, he was responsible for investigating the cause of the 1986 Space Shuttle Challenger disaster.

Feynman did pioneer work on quantum computing, the application of quantum mechanics to computing, and on nanotechnology. He latterly held the professorship in theoretical physics at the California Institute of Technology.

He was the author of many books, including the *"Surely You're Joking, Mr. Feynman!"* I mentioned earlier. His lectures were legendary, and he was loved by his students. Luckily, for those interested, many of his lectures can still be found online.

He died on 15 February 1988, aged just 69 in Los Angeles, USA.

And one the world ignored...

Rosalind Franklin

I have briefly discussed Einstein, Dirac, and Feynman as my favourite scientists. That is true, but I cannot close without telling you about one who, in my opinion, was unfairly treated and did not get the recognition she deserved. I also want to right an injustice that I believe was led primarily by the fact that she was a woman.

Rosalind Franklin

Rosalind Franklin was born on 25 July 1920 in Notting Hill, London. She had decided, by the time that she was 15 that she wanted to be a scientist. This was contrary to the norms of those days which expected girls to have an education fit for ladies (home economics, cooking, deportment, and so on, I suppose). Her father was opposed to her plans, which he did not consider appropriate. But Rosalind was a determined girl and stuck to her goals, and after an education at a London school for young ladies, went on to do a degree in Natural Sciences from Newnham College, Cambridge.

She started work in 1951 at Kings College in London and was immediately beset by issues raised solely because of her gender. She could not enter the senior common room where her male colleagues socialized and ate their lunch. Her accent (upper-class) was mocked. It was pure misogyny and discrimination. But she persevered.

Her field of study was chemistry. In her profession she became a specialist in X-ray crystallography, which is the use of X-rays to study the 3-dimensional shapes of molecules. She moved on from studying carbon and coal to biological molecules, such as viruses and, crucially, DNA. She successfully worked out the shape of the first virus to be discovered, which was called the Tobacco Mosaic Virus. Her conclusion about its shape was confirmed, but unfortunately, only after her death.

In the 1950s, James Watson and Francis Crick were trying to discover the structure of the DNA molecule. In parallel, though independently, Franklin was also working on the same topic, at Kings College, London, but using X-ray techniques. Watson and Crick's approach was to see what other scientists had achieved and to build on their published work. They also discussed with Franklin what she was doing.

Franklin and her PhD student, Raymond Gosling, had taken an X-ray photograph of the DNA molecule that is now famous, called Photo 51. Franklin's work showed that the structure of the

221

molecule was a double helix. Maurice Wilkins, a colleague of Franklin's at Kings College, showed the photo to Watson and Crick without Franklin's knowledge. It gave the two Cambridge researchers the key information that they lacked in completing the DNA structure.

Watson and Crick published their paper in *Nature* in 1953, but without acknowledging Franklin. In the same publication, Franklin and Wilkins also published their individual papers on the same subject. Watson and Crick were awarded the Nobel Prize in 1962, together with Wilkins, for discovering the shape of the DNA molecule.

Franklin had died of cancer in 1958, aged just 38 in Chelsea, London, well before the Nobel Prize was awarded in 1962. The Nobel Prize committee has a rule that prohibits Nobel Prizes being awarded posthumously. However, this rule only came into being in 1974! There is another Nobel Prize stipulation, that not more than three persons can share a prize, but of course it does not stipulate which prize someone can get. A viable alternative solution, which has already been suggested by others, is that two prizes could have been awarded, one to Watson and Crick for Physiology or Medicine and one to Franklin and Wilkins, for Chemistry.

Concerns remain about why Franklin's major contribution was ignored.

Rosalind Franklin has not been forgotten though. Many countries have issued stamps with her photo and the UK minted in 2020, the centenary of her birth, a special commemorative 50p coin bearing the famous Photo 51 image. ESA, the European Space Agency, together with the Russian Roscosmos State Corporation, is planning to send a robotic rover to Mars in 2022. The rover has been named "Rosalind Franklin".

Finally...

I started writing this book with an idea of leaving a legacy for my grandchildren. Then I thought, perhaps others may be interested in it as well. I find the whole subject of science, particularly astrophysics and cosmology, so enthralling, that I am the last person who should judge how true this is. However, I have given it a shot.

What I can tell, with confidence, is that I have thoroughly enjoyed the writing of it. For me, to coin a cliche, it genuinely has been a labour of love. It is now over 8 years since I started the project, way back in September 2012. The writing is done, but I am sure, the copyright checks and redrafts will take their own time. The enforced lockdown that the world has suffered in 2020, due to the coronavirus wreaking havoc to our normal lives, has perversely given me (and I am sure many others) the impetus to get on and finish what I had long promised myself to do. This book is therefore the consequence of my promise to myself.

I hope the book has given some explanation for the awe that many, I amongst them, feel when faced with the scale and grandeur of the Universe. Almost everything about it overwhelms the mind. I have tried to explain the underlying basics of what makes the cosmos tick. I hope that my readers, young as well as the mature have come to understand it a little better, and to appreciate more the human genius that has enabled us to unravel the mysteries.

If you come across errors in the dissertation, they are entirely mine. The truths in here are the discoveries of others, the scientists I have mentioned and many more besides. I have just been the storyteller.

If you have been intrigued enough to want to learn more about the topics covered in this book, I will encourage you to explore the wealth of literature on the subject that is readily available, both in physical book form and in the virtual world of the internet.

Good learning.

A special note for my younger readers

You, the young men, and women who read this book, have been very much in my thoughts as I have been writing. I hope that you have been fascinated by our Universe and its story and agree that it is utterly amazing. I also hope that you noted the references to the many scientists that I made along the way. It is the work of these, and many others not named, whose efforts enabled this story to come together. We hear and read of their achievements and marvel. But we often do not appreciate the hard work and dedication that has taken place in the background: the years of study that have gone into gaining the qualifications, and the research and experimentation that has produced the results. As Thomas Edison (1847-1931), the US inventor of the light bulb said, "Genius is one percent inspiration and 99% perspiration".

Working in science is like building a Lego structure. You put new bricks on what existed before. Sometimes you dismantle a part of what exists and add something new in its place. Slowly the solution develops, built on what has gone before. Once, Newton is said to have remarked: "If I have seen further, it is by standing on the shoulders of giants". So, if you are inspired enough to want to take up a career in astrophysics, cosmology, or any of the other STEM (Science/ Technology/Engineering/Mathematics) subjects, as I hope at least some of you will be, remember it will take hard work, but it will also be a source of much pleasure once you get into it. If you put in the effort, you will find a rewarding career and an exciting and fascinating future awaiting you, as well as having a real opportunity to make yourself, your country, and the world proud. Treat this book as a 'taster'. If you like the taste, go for the main course.

One of the other things you will have noticed as you went through the book is that virtually all the scientists referenced were men. In my view, this has nothing to do with gender ability, rather it is cultural, a reflection of the bias and prejudice about what women can/cannot, should/ should-not do. There have been brilliant women scientists in many disciplines, but nowhere near as many as should (or could) have been. We are making progress towards equality of opportunity and things are changing, but only slowly. This needs to speed up. We cannot afford to waste talent. We have so many unanswered questions that remain to be addressed, and so many complex problems, such as climate change, to be resolved.

We need you young people of today to grow up to think freely, to achieve great things and, on your own journey to greatness, to break down barriers that hinder the right of all to be represented on the global stage, irrespective of gender, race, creed, or other pointless prejudices.

Here's power to your elbows.

Athar Shareef

atharshareef@aol.com

10 December 2020

Image credits

Notes

1. Attributions are given below as follows: Fig. number in text; Title; Credit: Organisation(s)/ Person(s), changes by author (if any) / Licence (if any); URL for the image website.
2. Creative Commons licence references in the citations below are as follows:

   CC BY 2.0 refers to Creative Commons Attribution 2.0 Generic licence
   https://creativecommons.org/licenses/by/2.0/
   CC BY 3.0 refers to Creative Commons Attribution 3.0 Unported
   https://creativecommons.org/licenses/by/3.0/
   CC BY-SA 3.0 refers to Creative Commons Attribution ShareAlike 3.0 Unported
   https://creativecommons.org/licenses/by-sa/3.0/
   CC BY 4.0 refers to Creative Commons Attribution 4.0 International
   https://creativecommons.org/licenses/by/4.0/
   CC BY-SA 4.0 refers to Creative Commons Attribution ShareAlike 4.0 International
   https://creativecommons.org/licenses/by-sa/4.0/

3. The use of images in this book does not imply the endorsement of the book by those holding copyright in the images nor by those connected to them
4. The images in this book were accessed during September 2020 - January 2021
5. The images credited to Bryn Reade (www.smallhillproductions.com) are used under a license granted by him to the author.
6. All illustrations in the text other than those attributed below are the work and copyright of the author.

Attributions

Cover New view of the Pillars of Creation - visible; Credit: NASA, ESA/Hubble and the Hubble Heritage Team / CC BY 4.0;
https://www.spacetelescope.org/images/heic1501a/

Chapter 1
Fig. 1.1, 1.2 Newton and Earth images used in non-digital illustrations created by author; Credit: Bryn Reade (www.smallhillproductions.com)
Fig. 1.6, 1.9, 1.12 Visible spectrum image used in non-digital illustrations created by author; Credit: Bryn Reade (www.smallhillproductions.com)

Fig. 1.10 Albert Einstein,1879-1955; Downloaded from: Library of Congress (LoC) Prints and Photographs Division Washington, D.C. 20540 USA; Rights Advisory: No known restrictions on publication. No renewal in Copyright office (checked 8/31/76);
https://www.loc.gov/pictures/item/2004671908/

Fig. 1.11 The galaxy next door; Courtesy of: NASA/JPL-Caltech;
https://www.nasa.gov/mission_pages/galex/pia15416.html

Fig. 1.13 Cartoon ambulance car(s) isolated on white background. 3D render; Credit: Illustrations 184240494 and Cartoon ambulance car(s) isolated on white background. 3D render; Credit: Illustrations 184240494 and 184240502 © Nerthuz | Dreamstime.com, images inverted by author;

https://www.dreamstime.com/cartoon-ambulance-car-isolated-cartoon-ambulance-car-isolated-white-background-d-render-image184240494

https://www.dreamstime.com/cartoon-ambulance-car-isolated-cartoon-ambulance-car-isolated-white-background-d-render-image184240502

Chapter 2

Fig. 2.3 Bell Labs Horn Antenna Crawford Hill NJ; Credit: Fabioj / CC BY-SA 3.0;
https://commons.wikimedia.org/wiki/File:Bell_Labs_Horn_Antenna_Crawford_Hill_NJ.jpg

Fig. 2.4(a) COBE CMB; Credit: NASA;
https://science.nasa.gov/missions/cobe

Fig. 2.4(b) WMAP CMB; Credit: NASA/WMAP Science Team;
https://map.gsfc.nasa.gov/media/101080/index.html

Fig. 2.4(c) Planck CMB; Credit: ESA and the Planck collaboration;
http://www.esa.int/ESA_Multimedia/Images/2013/04/Planck_CMB_black_background

Chapter 3

Fig. 3.2(a) PIA07901: Barred Spiral Galaxy NGC 1365; Courtesy of: NASA/JPL-Caltech/SSC;
https://photojournal.jpl.nasa.gov/catalog/PIA07901

Fig. 3.2(b) Giant elliptical galaxy NGC 1316 in Fornax cluster; Credit: ESO / CC BY 4.0;
https://www.eso.org/public/images/eso0024a/

Fig. 3.2(c) A spattering of blue; Credit: ESA/Hubble, NASA, D. Calzetti (UMass) and the LEGUS Team / CC BY 4.0;
https://www.spacetelescope.org/images/potw1436a/

Fig. 3.3 Pleiades, NGC 1432/35, M45; Credit: NASA, ESA, AURA/Caltech;
https://hubblesite.org/contents/media/images/2004/20/1562-Image.html?news=true

Fig. 3.4 Milky Way and our location; Courtesy of: NASA/Adler/U.Chicago/Wesleyan/JPL-Caltech, annotated by author;
https://www.nasa.gov/mission_pages/sunearth/news/gallery/galaxy-location.html

Image credits

Fig. 3.11 Orion Night Shy Star Constellation Gray Stars; Credit: Image by sl1990 from Pixabay;
https://pixabay.com/photos/orion-night-shy-star-constellation-2942260/

Fig. 3.12 Cat's eye planetary nebula Ngc 6543 Planetary fog; Credit: Image by WikiImages from Pixabay;
https://pixabay.com/photos/cat-s-eye-nebula-ngc-6543-11162/

Fig. 3.13 Image of the Crab nebula (M1) taken from the earthbound Liverpool Telescope on La Palma; Credit: Göran Nilsson & The Liverpool Telescope / CC BY-SA 4.0, via Wikipedia Commons;
https://commons.wikimedia.org/wiki/File:The_Crab_Nebula_M1_Goran_Nilsson_%26_The_Liverpool_Telescope.jpg

Chapter 4

Fig. 4.3 First image of a Black Hole; Credit: EHT Collaboration / CC BY 4.0;
https://www.eso.org/public/images/eso1907a

Fig. 4.5 A rose made of galaxies; Credit: NASA, ESA and the Hubble Heritage Team (STSci/AURA);
https://spacetelescope.org/images/heic1107a/

Chapter 5

Fig. 5.1 New view of the Pillars of Creation – visible; Credit: NASA/ESA and the Hubble Heritage Team / CC BY 4.0;
https://www.spacetelescope.org/images/heic1501a/

Fig. 5.2 Formation of the Solar System, Birth of Worlds; Credit: NASA;
https://mobile.arc.nasa.gov/public/iexplore/missions/pages/yss/november.html

Fig. 5.4(a) Solar Eclipse 2017 Totality 2017 Tennessee Sun; Credit: Image by Jan Haerer from Pixabay, annotated by author;
https://pixabay.com/photos/solar-eclipse-2017-totality-2017-2670351/

Fig. 5.4(b) Realistic sun or star closeup 3D rendering illustration; Credit: iStock.com/libre de droit, cropped by author;
https://www.istockphoto.com/photo/realistic-sun-or-star-closeup-3d-rendering-illustration-gm1267178422-371712107

Fig. 5.5 Largest sunspot of the solar cycle; Courtesy of: NASA/SDO and the AIA, EVE, and HMI science teams;
https://www.nasa.gov/content/goddard/largest-sunspot-of-solar-cycle

Fig. 5.6 Magnificent CME erupts on the Sun with Earth to Scale; Courtesy of: NASA/SDO and the AIA, EVE, and HMI science teams;
https://images.nasa.gov/details-GSFC_20171208_Archive_e001660

Fig. 5.7 Sizes of the planets compared to the Sun; Embedded planet images Credit: NASA;
https://solarsystem.nasa.gov

Fig. 5.9 Terrestrial planet sizes; Credit: NASA/Lunar and Planetary Institute;
https://solarsystem.nasa.gov/resources/687/terrestrial-planet-sizes/

Fig. 5.11 Moon Crater Aerial View (2010); Credit: Shane.torgerson / CC BY 3.0;
https://commons.wikimedia.org/wiki/File:Meteorcrater.jpg

Fig. 5.12 Iceland Aurora Borealis Northern Lights Beautiful; Credit: Image by David Mark from Pixabay;
https://pixabay.com/photos/iceland-aurora-borealis-2111811

Fig. 5.13 7 Major Tectonic Plates: The World's Largest Plate Tectonic; Credit: Earth How, 7 Major Tectonic Plates: The World's Largest Plate Tectonics [Infographic];
https://earthhow.com/7-major-tectonic-plates/

Fig. 5.14 Ring of Fire; Credit: US Geological Survey;
https://pubs.usgs.gov/gip/dynamic/fire.html

Fig. 5.16 Laurasia-Gondwana; Credit: Lennart Kudling / CC BY 3.0;
https://commons.wikimedia.org/wiki/File:Laurasia-Gondwana.svg

Fig. 5.17 Moon; Credit: NASA;
https://solarsystem.nasa.gov

Fig. 5.18 Earthrise; Credit: NASA;
https://www.nasa.gov/multimedia/imagegallery/image_feature_1249.html

Fig. 5.19 Dark, Recurring Streaks on Walls of Garni Crater on Mars; Courtesy of: NASA/ JPL-Caltech/Univ. of Arizona;
https://www.nasa.gov/image-feature/jpl/pia19917/dark-recurring-streaks-on-walls-of-garni-crater

Fig. 5.20 PIA01483: Voyager Tour Montage; Credit: NASA/JPL;
https://photojournal.jpl.nasa.gov/catalog/PIA01483

Fig. 5.21 Jupiter Planet Galilean Monde Io Europa Ganymede; Credit: Image by WikiImages from Pixabay;
https://pixabay.com/photos/jupiter-planet-galilean-monde-io-11615/

Figs. 5.22 The Day the Earth Smiled; Courtesy of: NASA/JPL-Caltech/Space Science Institute;
https://www.nasa.gov/mission_pages/cassini/multimedia/pia17171.html

Fig. 5.23(a) Hazy Orange Orb; Courtesy of: NASA/JPL-Caltech/Space Science Institute;
https://images.nasa.gov/details-PIA14602

Fig. 5.23(b) Deepest Ever Dive Through Enceladus Plume Completed. Artist's concept; Courtesy of: NASA/JPL-Caltech;
https://www.nasa.gov/feature/jpl/deepest-ever-dive-through-enceladus-plume-completed

Fig. 5.23(c) Iapetus; Courtesy of: NASA/JPL-Caltech/Space Science Institute;
https://solarsystem.nasa.gov/moons/saturn-moons/iapetus/in-depth/

Image credits

Fig. 5.24 Kuiper Belt; Credit: NASA, annotated by author;
https://solarsystem.nasa.gov/solar-system/kuiper-belt/overview/
Fig. 5.25 PIA11709: A Colorful 'Landing' on Pluto; Credit: NASA/John Hopkins University Applied Physics Laboratory/ Southwest Research Institute;
https://photojournal.jpl.nasa.gov/catalog/PIA11709
Fig. 5.26 The Oort Cloud; Credit: NASA;
https://solarsystem.nasa.gov/resources/491/oort-cloud/?category-solar-system_oort-cloud

Chapter 6
Fig. 6.2 Hamelin Pool (Stromatolites); Credit: Phil Whitehouse from London, United Kingdom / CC BY 2.0;
https://commons.wikimedia.org/wiki/File:Hamelin_Pool_(Stromatolites)_(2051681803).jpg
Fig. 6.3 A deep-sea hydrothermal vent; Credit: Oregon State University / CC BY-SA 2.0;
http://www.sci-news.com/biology/life-hydrothermal-vents-07772.html
Fig. 6.4 5 ways to search for exoplanets; Credit: Adapted by author from: NASA "5 Ways to Find a planet";
https://exoplanets.nasa.gov/alien-worlds/ways-to-find-a-planet/#
Fig. 6.5 The "Wow!" signal; Credit: Big Ear Radio Observatory and North American AstroPhysical Observatory (NAAPO);
http://www.bigear.org/Wow30th/wow30th.htm
Fig. 6.6 PIA22835: Two Interstellar Travelers; Courtesy of: NASA/JPL-Caltech;
https://photojournal.jpl.nasa.gov/catalog/PIA22835

Appendix A
Fig. A.8 Representation of the Earth affecting the curvature of the Space-Time graph; Credit: By Mysid, made in Blender & Inkscape, CC BY-SA 3.0;
https://commons.wikimedia.org/w/index.php?curid=45121761
Fig. A.12 Hubble captures a "lucky" galaxy alignment; Credit: ESA/Hubble and NASA / CC BY 4.0, image cropped by author;
https://www.spacetelescope.org/images/potw1151a
Fig. A.13 LISA, space-borne gravitational wave detector; Credit: NASA/ ESA/GSFC/ASD;
https://lisa.nasa.gov/archive2011/

www.ingramcontent.com/pod-product-compliance
Lightning Source LLC
Chambersburg PA
CBHW050037220326
41599CB00040B/7190